An informal intro to gauge field theories

IAN J. R. AITCHISON

Lecturer in Theoretical Physics, University of Oxford

The right of the University of Cambridge to print and sell all manner of books was granted by Henry VIII in 1534. The University has printed and published continuously since 1584.

CAMBRIDGE UNIVERSITY PRESS
Cambridge
London New York New Rochelle
Melbourne Sydney

CAMBRIDGE UNIVERSITY PRESS
Cambridge, New York, Melbourne, Madrid, Cape Town, Singapore, São Paulo

Cambridge University Press
The Edinburgh Building, Cambridge CB2 8RU, UK

Published in the United States of America by Cambridge University Press, New York

www.cambridge.org
Information on this title: www.cambridge.org/9780521245401

First published 1982
Reprinted with corrections 1984
This digitally printed version (with additional corrections) 2007

A catalogue record for this publication is available from the British Library

Library of Congress Catalogue Card Number: 81–21753

ISBN 978-0-521-24540-1 hardback
ISBN 978-0-521-03954-3 paperback

An informal introduction to gauge field theories

Contents

Preface

Relativistic quantum field theory is firmly re-established as the most successful and fruitful general framework for the study of the fundamental interactions. Within this framework, the class of theories called *gauge field theories* is of quite special interest. The currently favoured theories of the strong and weak interactions are of this type, as is the well-established theory of the electromagnetic interaction, quantum electrodynamics (QED). Elementary particle physics is now almost exclusively concerned with such theories.

Quantum field theory, however, can easily be presented as a highly formal - and formidable - discipline; and many of the technical complexities are particularly evident in the case of gauge field theories. In such a situation one can espouse from the start a formal approach which is broad enough to cope with all these complexities: but the essential mathematical and physical ideas may be somewhat obscured, and the path to practical calculations (that is, the Feynman rules) is long. Alternatively, one can short circuit the formalism and reach the calculational stage more quickly, by emphasising the important physics and relying on the use of analogy and heuristic argument where necessary. The latter strategy is followed in the book *Gauge Theories in Particle Physics* by A. J. G. Hey and myself, which is in some ways complementary to this one.

The present book is quite definitely *not* a treatise on field theory, nor even on gauge field theories. In the first place, it assumes that the reader is familiar with the rudiments of relativistic quantum field theory - via, say, a course which concentrated mainly on scalar fields. Canonical quantisation of such fields (and of spinor fields) will be assumed, and also the route to Feynman graphs from a simple Lagrangian, via Wick's theorem. It seems to me that there is merit in separating off such a 'first introduction' to field theory itself, from the study of gauge field theories in particular, since the latter do present special features and difficulties which, I believe, make their study inappropriate to a first course on field theory. On the more positive side, it is very natural to study gauge theories on their own, since - apart from the fact that they seem to be so successful - the extent to which they are all based on variations on a common theme of *symmetry* allows an integrated and unified treatment.

The treatment presented here is, of course, field theoretic, but (for the most part) *informal.* It aims to steer a middle course between the complete formal approach and the non-field-theoretic calculational emphasis of Aitchison & Hey. By 'informal' I mean several things: I mostly discuss simple, illustrative, but relevant, special cases rather than give general and abstract treatments; and I attempt to illuminate some of the most difficult, but important, points by emphasising physical ideas (such as unitarity, high energy behaviour, superfluidity, superconductivity, ...). I also think that the canonical approach is, still, less formal than the path-integral one.

This book is intended very much as an introduction to the subject, in which the emphasis is on the basic ideas of gauge theories. I have included relatively little in the way of practical application, and only dealt very briefly with some of the still developing areas of the subject. I have not been able to include any discussion of non-perturbative methods. I have, however, tried to explain, in a reasonably elementary way, some of the technical aspects of canonical quantisation of gauge fields. And in the final chapter the study of more advanced topics related to renormalisation is initiated: renormalisability is one of the prime characteristics of gauge theories, so that some discussion of it must be attempted, even in an introductory treatment. As with quantisation itself, particular subtleties are encountered in the renormalisation process.

As I have said, gauge theories are intimately related to symmetries. We now recognise two quite distinct forms of symmetry in quantum field theory - 'manifest' and 'hidden'. A further distinction, basic to gauge theories, is that between a 'local' and a 'global' symmetry. Local symmetry and its relation to dynamics is the central underlying theme of gauge field theory. The first main section of the book (Chapters 1 to 4) is devoted essentially to manifest symmetry, mainly of the local (gauge) kind. Chapter 5 forms a bridge to the second section (Chapters 6 and 7), which is concerned with hidden symmetry. In the final chapter, which constitutes a third section, the reader will find somewhat less detailed explanations than in the earlier chapters, and more appeal to outside sources for certain results; the aim here is to provide enough apparatus to allow reasonable understanding of some important but more technical aspects (which have been touched on earlier) - and also to facilitate access to more specialised texts.

The book is based on lectures given at the British Universities Summer School on Theoretical Particle Physics, held at the University of Manchester in September 1979. I am grateful to Dr G. Shaw and the BUSS Committee for the invitation to give a course on gauge theories. Much of my education in the subject took place during a year (1978-9) spent on leave at the University of Rochester. I am very grateful to the R. T. French Company for providing full financial support for that year. While at Rochester, I benefitted greatly by attending lectures given by Professor V. S. Mathur; and I enjoyed many discussions with him and other members of the particle group there. It is a pleasure to take this opportunity of thanking Professor H. E. Gove and the members of the Physics Department at Rochester for their generous hospitality. I have enjoyed many useful discussions with colleagues at Oxford, among them especially Dr C. M. Fraser, Dr R. L. Heimann, Dr C. H. Llewellyn Smith, Dr J. E. Paton and Dr G. G. Ross. Finally, I am very much indebted to Dr P. V. Landshoff, for carefully and critically reading both a preliminary and the final version.

Ian J. R. Aitchison

January 1982

Preface to the Paperback Edition

The first printing of this book in 1982 was followed by a reprint with corrections, which is the text reissued here. During the intervening year, 1983, the discoveries of the $W^{\pm}$ and Z^{0} were announced, as recorded in footnotes on pages 4 and 110. The discovery of the top quark came twelve years later, completing the third family. Of the particles expected in 1984, only the Higgs boson has so far eluded detection. Despite this notable absentee, there is – remarkably – still no conclusive quantitative discrepancy between experiment and the predictions of the Standard Model of particle physics, based on the gauge group SU(3) × SU(2) × U(1) and a single Higgs doublet, as outlined here.

The opportunity has been taken here to make some further corrections for the paperback reissue. I am extremely grateful to Dr. George Emmons for essential help with this.

Ian J. R. Aitchison

February 2007

1 Introduction and motivations

The Glashow-Salam-Weinberg (GSW) theory of the electroweak interactions, and the theory of the strong interactions between quarks based on colour-dependent forces (quantum chromodynamics, or QCD), are both gauge field theories - remarkable generalisations of the first gauge field theory, quantum electrodynamics (QED). QED is a supremely successful theory; the GSW theory now also has an impressive amount of empirical support; and there is a substantial amount of qualitative - and, increasingly, of quantitative - empirical evidence, coupled with strong theoretical arguments, in favour of QCD. In the following section we shall review, very briefly, some features of strong and weak interactions which provide empirical motivations for these theories - and which might seem reasons enough for studying them. On the other hand, it is always interesting to enquire why a theory *had* to be the way it is, and whether it could not have been somewhat different: to discover, in fact, some *a priori* principles which might severely constrain, or even perhaps determine, the form of the interactions. In Section 1.2 we shall introduce the idea (see, for example, Weinberg 1980) that renormalisability and symmetries might usefully be considered as such principles. Such arguments will inevitably not be conclusive: nevertheless, it seems pedagogically preferable to provide some initial motivation for gauge theories, since such theories are, if beautiful in conception, technically very complicated in practice. The reader should at once be assured that several concepts to be mentioned below, which may well be unfamiliar, will be discussed much more fully in later chapters.

1.1 Empirical motivations

We begin with QCD. There is now ample spectroscopic evidence that hadrons are composites, made of quarks, denoted by q, and antiquarks ($\bar{q}$). In addition to their *flavour* degrees of freedom (isospin, strangeness, charm, ...), quarks carry a further degree of freedom, called *colour*, which is three-valued ('red', 'green', or 'blue'). Observed hadrons have no net colour, being analogous to neutral atoms. The forces between quarks must be colour-dependent, otherwise, for every 'colourless' meson in a $q\bar{q}$ configuration, for example, there would be eight other coloured ones degenerate with it, and these are not observed. The forces are presumably mediated by field quanta, which are called gluons and carry colour quantum numbers. Indeed, there is evidence from deep inelastic scattering of e, μ and ν from nucleons that roughly 50% of a rapidly moving nucleon's momentum is carried by particles that are neutral with respect to the electromagnetic and weak forces; these particles are identified with gluons. The gluons should have odd spin, since even spin field quanta couple in the same way to particles and antiparticles, and hence would lead to unwanted qq states degenerate with the $q\bar{q}$ mesons. The simplest possibility is $J = 1$. This has the further point in its favour that, in the limit in which the quark masses go to zero, a theory with vector-mediated forces exhibits chiral symmetry (see Section 6.8 below), and there is good evidence for approximate chiral symmetry (realised in the 'hidden' mode, see Chapter 6) in hadronic physics.

As regards the colour dependence of the interquark force, the extraordinary 'saturation' property (qqq and $q\bar{q}$ states exist, but $q\bar{q}q$ and $qqq\bar{q}$, for example, do not) can be qualitatively understood by assuming that colour is associated with an $SU(3)$ symmetry group, called $SU(3)_C$. If quarks belong to a triplet, and gluons to an octet, representation of $SU(3)_C$, the $SU(3)_C$-invariant, one-gluon exchange force between quarks will result in colour singlet states qqq and qq having lower energy than any colour non-singlet state (such as $q\bar{q}q$). Further, the short-range spin-dependent force abstracted from one gluon exchange gives a remarkably successful simple explanation of the baryon mass splittings (Close 1979, Isgur & Karl 1979).

This theory - triplets of coloured quarks interacting in an $SU(3)_C$-invariant way by the exchange of octets of coloured gluons - is not yet QCD. To obtain QCD, one further piece of empirical evidence can be added: the quarks in a nucleon, when probed in deep inelastic lepton scattering, behave as if they are essentially free (for a review, see Close 1979). This implies that the interquark force must become weak at short distances; it may vanish as the interquark separation tends to zero. This property is known as 'asymptotic freedom' (see Section 8.4 below), and is remarkable in being possessed by only a certain class of theories - namely, non-Abelian gauge field theories (Politzer 1973, Gross & Wilczek 1973*a*). Much of the following chapters will, of course, be concerned with explaining what these theories are: for the moment we simply state that QCD is a sort of generalised quantum electrodynamics (QED), in which the eight gluons play a role analogous to that of the single photon, and in which the gauge-invariance of QED is generalised to a more complicated non-Abelian gauge-invariance (to be further discussed in the following section, and fully in Chapter 3) involving the colour group $SU(3)_C$.

The above is no more than a very brief outline of some salient aspects of strong interaction physics, which point towards QCD. Much work has been done, and remains still to be done, in effecting detailed comparisons between this theory and experiment. Progress is hampered by the very fact that QCD is a theory of the *strong* interactions: the extraordinary predictive power of QED relies entirely on the use of perturbation theory, which will generally not be applicable for strong interactions. However, the asymptotic freedom property encourages hope that, at least in certain kinematic limits, perturbative QCD may be applicable to experiment. It would be inappropriate here to attempt a more detailed review of the present empirical status of perturbative QCD, which is already a very large and still expanding field of activity. Progress can also be expected, in the future, from the development of non-perturbative methods.

We turn now to weak interactions, which are viewed as occurring, at the present level of elementarity, among leptons and quarks. There is certainly a great deal of experimental information on weak interactions - but, for a reason which will appear shortly, it cannot yet lead us directly to the GSW theory, despite some extremely encouraging areas of agreement between that theory and experiment (and the absence of any clear disagreements, so far).

A major qualitative *prediction* of the GSW theory was the existence of 'weak neutral-current' processes. The ordinary Fermi-type processes such as $n \to pe^-\bar{\nu}_e$ or $\nu_\mu + p \to \mu^- +$ anything, had long been known: in these 'charged current' processes, the lepton pair always carries one unit of charge. In 1973 (Hasert et al. 1973; also Benvenuti et al. 1974) it was established that processes such as $\nu_\mu + p \to \nu_\mu +$ anything, also occur, in which the leptons are neutral. The mere existence of such processes is, of course, only a necessary - not sufficient - condition for the GSW theory to be true; many other theories with neutral currents are possible.

The next important feature of weak interactions we consider, is their range. In Fermi's original theory (Fermi 1934*a*, 1934*b*) the β decay $n \to pe^-\bar{\nu}_e$ was assumed to be of *zero* range; and there is still no evidence of any finite range for any weak interaction. Fermi's theory is said, alternatively, to be 'pointlike': the process in which the neutron turns into a proton, and the $e^-\bar{\nu}_e$ pair is emitted, takes place at a single space-time point. In a Lagrangian, such an interaction would involve the product of the fields for the four participating fermions, all evaluated at the same space-time point: it is a 'four-fermion' interaction. This type of interaction differs fundamentally from Yukawa's picture of forces as being due to the exchange of quanta. In the Yukawa picture, the neutron would turn into a proton emitting a weak quantum (W^-) which would then - at a different space-time point - materialise as the $e^-\bar{\nu}_e$ pair. Other charged current processes would involve the analogous W^+, and neutral-current ones a neutral quantum Z^0. In the usual way, the distance over which such quanta can propagate - i.e. the range of the weak force - is inversely proportional to the mass of the quantum. Experimental upper limits on the range of the interaction correspond to lower bounds on the masses of such quanta - if they exist.

As is usual in quantum mechanics, the short-distance behaviour of interactions is probed by doing experiments at increasingly high energies. Nowadays, lepton beams are available with energies far greater than those released in weak decay processes (such as a β decay). Evidence for a non-zero range of the weak force can be obtained by looking for deviations from the 'pointlike' behaviour predicted, according to the four-fermion model, for neutrino-induced reactions (much as evidence for finite nuclear size was deduced by comparing scattering data with the pointlike Rutherford cross-section). In terms of the Yukawa picture, the absence of such deviations in the present data provides a lower bound on the masses of $W^{\pm}$ weak quanta, which is in the region of 25-30 GeV.

Of course, the direct production of massive weakly interacting quanta (further properties of which will be discussed shortly) should also be possible, if they exist. As we shall see in due course, the GSW theory involves just such massive weak quanta (the W^+, W^- and Z^0), with masses in the range 80-90 GeV. But, at the time of writing, no accelerator is capable of producing such heavy particles (the first machine to have the capability will be the CERN Collider; subsequently, LEP should be ideal for the purpose). Present experiments are therefore, on this scale, all at 'low energies'; they cannot yet lead us directly to the full GSW theory, which appears effectively pointlike at energies available in 1981–2.†

Much is, however, known about the couplings in the (presumably effective) four-fermion interactions. In Fermi's original theory, the charged lepton pair, emitted at a point in the $n \to p$ transition, was regarded as analogous to the photon emitted in an electromagnetic transition. The analogy led to the idea of 'weak vector currents': the $n \to p$ transition was described by the hadronic weak current matrix element $\langle p|J_\mu|n\rangle$, while the $e^-\bar{\nu}_e$ pair were also coupled in a leptonic weak current form $\langle e^-\bar{\nu}_e|j_\mu|0\rangle$. The matrix element was then proportional to the (Lorentz invariant) product of these two currents - the so-called 'current-current' interaction. More than twenty years after Fermi's papers, it was established that these charged current matrix elements are indeed of vector type, but not pure vector - rather, the famous 'V–A' *mixture* of vector and axial vector, incorporating the observed parity violation effects. The neutral current interactions are also known to be vector in character and parity violat-

† The discovery of the $W^{\pm}$ and Z^0 particles was reported in 1983 by two groups at the CERN $\bar{p}p$ Collider. UA1 collaboration: ($W^{\pm}$) Arnison et al. (1983*a*), (Z^0) Arnison et al. (1983*b*); UA2 collaboration: ($W^{\pm}$) Banner et al. (1983), (Z^0) Bagnaia et al. (1983).

ing although they are *not* of precisely the same V–A structure as the charged currents.

We note that the analogy with electromagnetism is much improved if the Yukawa picture is adopted. The mediating quanta $W^{\pm}$ and Z^0, not the lepton pairs, are then the analogues of the photon, and the vector character of the interactions implies that, like the photon, they have spin one.

Experiment can also determine the strengths of the various transition matrix elements (or couplings in the Yukawa picture). Instead of the potentially enormous number of *a priori* independent strength parameters, some remarkable patterns have emerged. We consider the charged current (CC) couplings first. All purely leptonic CC interactions seem to have the same overall strength, which may be taken to be that appearing in the current-current form of the $\mu^- \to e^- \bar{\nu}_e \nu_\mu$ matrix element. As regards the CC interactions of hadrons, the same universal strength appears to govern weak interactions among all quarks, after allowance is made for Cabibbo mixing (and its generalisations) among quark flavours (see Sections 7.5 and 7.6 below). Finally, present data on neutral current processes show that they can all be described in terms of just one additional parameter.

The remarkable feature of 'universality of coupling', for quarks as well as for leptons, clearly implies the existence of some powerful underlying *symmetries*, and associated conservation laws. Early evidence of such regularities was interpreted by Feynman & Gell-Mann (1958) as indicating that the charge-changing vector currents were conserved, like the electromagnetic one – we shall discuss why in Section 8.2. Associated with these conserved currents were 'weak charges', interpreted by Feynman & Gell-Mann as being proportional to the isospin raising and lowering operators of an $SU(2)$ symmetry group. The universality of coupling was also perceived as characteristic of a *gauge* theory (as will be explained in Section 3.3), and several attempts were made to find a scheme which comprehended the emergent, but elusive, analogy between weak and electromagnetic interactions. It was Glashow (1961) who suggested the $SU(2) \times U(1)$ gauge structure, which now appears to describe correctly the symmetry properties of electroweak phenomena. A remaining theoretical stumbling-block was the problem of understanding how the weak vector bosons could be massive, while still being (like the photon) gauge field quanta. The phenomenon of 'hidden' gauge-invariance resolved this dilemma, and was postulated in the models of Weinberg (1967) and Salam (1968). In the low energy (current-current) form of their scheme, which uses the Glashow gauge group, NC interactions are indeed – as we shall see in Chapter 7 – determined in terms of CC ones and one further parameter, θ_W.

The work of submitting the GSW theory – even in its low energy form – to detailed experimental scrutiny continues, and will not be further reviewed here.

Before long the more complete high energy structure, involving the intermediate vector bosons and their couplings, should also begin to be unravelled. Meanwhile, further - and independent - motivation for the theory may be provided by non-empirical arguments, which we now consider.

1.2 Theoretical motivations

Contemporary theories of the fundamental interactions take for granted that the appropriate general framework is provided by the Lagrangian formalism of relativistic quantum field theory (briefly described in the following chapter). A major difficulty of such theories is that, when calculated beyond the lowest-order approximation, ultraviolet divergences almost invariably appear (see Section 8.1). Such infinities were first noticed in QED (then recently developed) by Oppenheimer (1930) and Waller (1930*a*, 1930*b*). Almost twenty years later, Feynman, Schwinger, Tomonaga, and Dyson, showed in an independent series of papers (collected in Schwinger 1958) how all the infinities of QED could be eliminated if the observed (finite) values of the electron mass and charge were interpreted, not as the parameters m and e appearing in the original Lagrangian which defined the perturbation theory, but as those 'renormalised' quantities which resulted when m and e were corrected by all higher-order effects due to virtual photons and e^+e^- pairs. Taking these two parameters from experiment, spectacular agreement with many predictions was, and still is, obtained. An introduction to the renormalisation procedure is given in Section 8.1.

It was soon realised that only for a limited class of field theories was renormalisation possible: for them, finite results for all predictions could be obtained by taking a finite number of parameters from experiment. But for the other, *non-renormalisable* theories, this was not possible - in these terms, an infinite number of parameters would have to be supplied. In particular, the four-fermion interaction of the original Fermi theory of weak interactions was recognised as non-renormalisable. Thus renormalisability becomes a useful principle, *constraining* possible theories to belong to that class for which finite calculations in perturbation theory, using a finite number of input parameters, are possible.

We have seen that there are good empirical reasons for thinking that, in terms of a Yukawa picture, both the strong and the weak interactions are mediated by vector quanta, as are the electromagnetic ones. It turns out that the question of the renormalisability of all such vector theories is particularly intricate. In the case of QED, the *gauge-invariance* of the theory plays a crucial role in the proof of renormalisability (see Section 8.2 below). For QCD, the requirement of renormalisability can be substituted for that of asymptotic freedom, in our discussion in the preceding section: the $SU(3)_C$ symmetry implied by aspects of the data must be turned into a generalised (*non-Abelian*) gauge symmetry (see below, and Chapter 3) if the theory is to be renormalisable. Asymptotic freedom then becomes a prediction (Section 8.4).

For both QED and QCD, the gauge-invariance is associated with the masslessness of the vector quanta. The weak interaction case is more subtle. We saw that phenomenology led us to a theory of massive charged vector type. Now such theories will *not* normally exhibit any gauge-invariance symmetry (see Section 3.2); nor, as we shall indicate in Section 5.2, will they be renormalisable. Remarkably enough, however, we shall be led, in Section 5.2, to the idea that one particular type of massive vector theory *is* renormalisable - namely, a theory with a 'hidden' gauge symmetry. This is an absolutely crucial concept, which we shall explain in the rest of that, and the following, chapter. The application of these ideas to weak interactions will be discussed in Chapter 7.

Renormalisability can be used as a powerful guide in theory construction. On the other hand, we seem to have arrived at the position that the interesting theories are all *gauge theories* - yet these theories do not exhaust the class of renormalisable field theories. We may still ask 'Why gauge theories?' A possible clue is provided by the essential role played by *symmetry* in such theories.

The idea that it is possible to find a symmetry principle powerful enough to dictate the form of the interactions seems to go back to Einstein and Minkowski, originating in their work on electromagnetism. The Maxwell equations were discovered as a synthesis of the experimentally established laws of electromagnetism, and the equations then revealed a symmetry - Lorentz invariance. Minkowski developed the tensor calculus of special relativity, which provides the mathematical tools and concepts for analysing whether a given theory is covariant under Lorentz transformations or not: the symmetry principle that all inertial frames be equivalent becomes the mathematical requirement that physical laws be covariant under Lorentz transformations, which is implemented immediately using Minkowski's tensor calculus. Of course, this requirement, though restrictive, by no means fixes the forms of the interactions. In seeking to *enlarge* the invariance group of special relativity, Einstein was led to study general coordinate transformations. The requirement of general covariance, when combined with the principle of equivalence, did indeed lead to a new theory of a specific interaction, namely, gravity (though one has to note that the field equations are not uniquely determined by the covariance requirement alone).

Reversing the line of thought, one can re-interpret this profoundly novel approach to a theory of interaction by saying that the existence of the gravitational field 'allows' the very wide class of general coordinate transformations to function as a symmetry group. It is then natural to ask whether the electromagnetic field can be 'explained' in terms of some further symmetry which is made possible by its existence. For many years Einstein tried to enlarge the transformation group of general relativity so as to include electromagnetism in a *unified* theory with gravitation, but without success. It was Weyl who introduced the particular enlargement which became known as 'gauge-invariance'. He considered scale transformations (the translation of his German word should perhaps better have been 'scale invariance') as a possible invariance of the world, and tried to interpret the four-vector potential A_μ as the field which made this invariance possible.

The scale transformations considered by Weyl were space-time dependent, being different at each point. There is an important distinction to be made between transformations of this type, which depend on the local space-time point, and those which are the same at all space-time points: the former are called *local* transformations, the latter *global* ones. It is the requirement of local invariance which seems to be powerful in 'generating' interactions. Einstein's theory of gravity also illustrates this, because it emerges when the Lorentz coordinate transformation (which, since the matrix of coefficients is space-time independent, is a global one) is generalised to arbitrary transformations, dependent on the local space-time point (Utiyama 1956).

Weyl's idea did not succeed in its original form, but it was re-born in a different guise some years later, after the advent of quantum mechanics. It was realised by Fock (1927), London (1927) and by Weyl himself (Weyl 1929), that quantum electrodynamics was indeed related to a new local symmetry - that in which the electron's wavefunction undergoes a phase change which varies from point to point in space-time; when this transformation is carried out on the wavefunction, invariance is restored by performing a corresponding gauge transformation on A_μ (as we shall see, in the field theory context, in Chapter 3). The scale transformations of Weyl's original idea had become *phase* transformations, but the words *'gauge transformation'*, and *'gauge-invariance'* for the associated symmetry, were retained. A more detailed review of the history of gauge theory is given by Yang (1977).

The simple local phase invariance exhibited by quantum electrodynamics is now called a $U(1)$ gauge symmetry, because a change of phase can be regarded as multiplication by a unitary matrix of dimension one. More complicated unitary transformations are of course possible, in which several wavefunctions (or fields) transform together: $SU(2)$ or $SU(3)$ transformations, for example, which involve unitary matrices acting on multiplets, can be thought of also as phase transformations, but ones in which the 'phase' is a matrix quantity. These are called non-Abelian symmetries, because the matrices will generally not commute. The original $SU(2)$ symmetry of Heisenberg, and its extension to $SU(3)$ and larger 'flavour' groups, were global non-Abelian symmetries: by themselves, they did not lead to a dynamical theory. But in a remarkable paper, Yang & Mills (1954) considered the local version of such symmetries, and thus initiated the study of non-Abelian gauge theories (see Chapter 3), which are beautiful generalisations of the (Abelian) $U(1)$ gauge theory which is QED.

The correct application of the ideas of Yang & Mills was neither immediate nor straightforward, as we shall see in later chapters; in particular, a crucial ingredient in modern gauge theories is the phenomenon of 'hidden' gauge-invariance (see Chapter 6), which was understood in the early 1960s. Its application by Weinberg and Salam led, as we saw above, to a renormalisable theory of weak interactions, and to the understanding of how the electromagnetic and weak forces could after all be unified, despite the fact that the quanta of the one are massless, but of the other are massive. With QCD also a gauge theory, there is a very strong suggestion that all three forces may be *unified* in one larger (local) group structure, encompassing the $SU(2) \times U(1)$ structure which partially unifies the electromagnetic and weak interactions; we shall comment briefly on this in Section 7.6. Such unification - the eventual hope of the analogies indicated in the previous section - is itself an argument for a gauge theory approach to other interactions, given the great success of QED. What of gravity itself? Einstein's grand vision of the eventual unification of the fundamental forces is being actively pursued again, but this time it is gravity which is, as it were, seeking unification with the other three. It is, however, premature to include in this book discussion of these attempts to find a 'Super Grand Unified Theory'.

It should be quite clear that *symmetry* is fundamental to all gauge theories, and consequently we now proceed to give an introduction to the subject of symmetries in quantum field theory.

2 Symmetries in quantum field theory: I Manifest

The purpose of this chapter is to collect together some simple ideas and results about symmetries, which are especially important for the understanding of gauge field theories. We assume an elementary knowledge of Lagrangian field theory, and use the conventions as to metric, etc., of Bjorken & Drell (1965). We defer discussion of the phenomenon of 'hidden' symmetry to Chapter 6.

2.1 The Euler-Lagrange equations: canonical quantisation

We consider systems described by a Lagrangian density $\mathcal{L}$, which depends on the fields and their gradients, and which has no explicit dependence on the space-time coordinates. The action I is the four-dimensional integral of $\mathcal{L}$:

$$I = \int d^4x \mathcal{L} \tag{2.1}$$

The integral extends over all space, and the fields will be assumed to vanish sufficiently fast at infinity to justify integration by parts. Lorentz covariance is assured if equations of motion (field equations) are derived from the action principle - that I be stationary under variation of the fields - and if $\mathcal{L}$ is a Lorentz invariant function of the fields and their gradients.

For simplicity, we begin with the case of a single scalar field $\phi(x)$, with $\mathcal{L}$ a function of $\phi(x)$ and $\partial_\mu\phi(x)$. Under an infinitesimal variation $\delta\phi(x)$, I changes by

$$\delta I = \int d^4x \left\{ \frac{\partial \mathcal{L}}{\partial \phi(x)} \delta\phi(x) + \frac{\partial \mathcal{L}}{\partial(\partial_\mu \phi(x))} \delta(\partial_\mu\phi(x)) \right\}. \tag{2.2}$$

Integrating by parts and using the relation

$$\delta(\partial_\mu\phi(x)) = \partial_\mu(\delta\phi(x)), \tag{2.3}$$

we obtain

$$\delta I = \int d^4x \left\{ \frac{\partial \mathcal{L}}{\partial \phi(x)} - \partial_\mu \frac{\partial \mathcal{L}}{\partial(\partial_\mu \phi(x))} \right\} \delta\phi(x). \tag{2.4}$$

Requiring this variation of I to vanish for arbitrary variations $\delta\phi(x)$ gives the Euler-Lagrange equations

$$\partial_\mu \frac{\partial \mathcal{L}}{\partial(\partial_\mu \phi(x))} - \frac{\partial \mathcal{L}}{\partial \phi(x)} = 0. \tag{2.5}$$

For example, if $\mathcal{L}$ has the simple form

$$\mathcal{L} = \tfrac{1}{2}[\partial_\mu \phi(x)\, \partial^\mu \phi(x) - m^2 \phi^2(x)], \tag{2.6}$$

(2.5) results in the Klein-Gordon equation

$$(\Box + m^2)\phi(x) = 0, \tag{2.7}$$

the field equation for a free scalar field of mass m. The above can be generalised straightforwardly to the case of several fields $\phi_r(x)$, of various Lorentz type (spinor, vector, etc.).

So far, nothing has been said about quantisation of the fields. Two approaches to quantisation are currently used, the 'canonical' procedure, and the path-integral method. The latter approach has been remarkably developed in recent years, and is now essential knowledge for the particle theorist. If offers a unified treatment of quantum mechanics, field theory, and statistical mechanical models and, moreover, appears particularly well suited to the difficult case of quantisation of the systems with constraints - and, as we shall see, gauge field theories are of just this type. However, our purpose is to present a more 'informal' introduction to the subject, and so we shall follow the traditional canonical procedure, omitting, furthermore, discussion of many of the difficulties and subtleties encountered. We can do this the more readily since several excellent books now exist which repair these deficiencies - specifically, those by Itzykson & Zuber (1980) (which describes both quantisation procedures) and Ramond (1981) (which provides a thoroughgoing path-integral approach *ab initio*).

The canonical procedure is a plausible generalisation, to systems with infinitely many degrees of freedom (fields), of the way in which the quantum mechanics of particles can be obtained from the corresponding classical mechanics - namely, by replacing classical observables by operators, and Poisson brackets by commutators (divided by i). The Lagrangian density $\mathcal{L}$ is used to define the 'momentum' $\pi_r(x)$ which is canonically conjugate to the field variable $\phi_r(x)$:

$$\pi_r(\mathbf{x}, t) = \frac{\partial \mathcal{L}(\phi, \partial_\mu \phi)}{\partial[\partial_0 \phi_r(\mathbf{x}, t)]}, \tag{2.8}$$

where we have indicated that $\mathcal{L}$ depends on ϕ and $\partial_\mu \phi$ only, and we also write in the time coordinate explicitly, to show that quantisation will be performed at a fixed time. To quantise, we replace the classical fields by operators satisfying the equal time commutation relations

$$[\phi_r(\mathbf{x}, t), \pi_s(\mathbf{x}', t)] = \mathrm{i}\delta_{rs}\delta^3(\mathbf{x} - \mathbf{x}') \tag{2.9}$$

$$[\pi_r(\mathbf{x}, t), \pi_s(\mathbf{x}', t)] = 0 \tag{2.10}$$

$$[\phi_r(\mathbf{x}, t), \phi_s(\mathbf{x}', t)] = 0. \tag{2.11}$$

For fermion fields, commutators are replaced by anticommutators.

2.2 Rotational invariance

We begin our discussion of symmetries with a reminder about a familiar example, that of invariance under a rotation of the space-time coordinates. It is as well to remember that, although for the rest of this book we shall be concerned with 'internal' symmetries, rather than with space-time ones, the two types of symmetry are formally handled in the same way; furthermore, the internal symmetries with which we shall deal, in the case of gauge fields, are actually space-time dependent anyway, as is to be expected from the 'geometrical' approach to dynamics hinted at in Section 1.2.

It is sufficient to consider an infinitesimal rotation, which we parametrise as

$$\mathbf{x}' = \mathbf{x} + \boldsymbol{\epsilon} \times \mathbf{x} \tag{2.12}$$

in terms of three parameters $\boldsymbol{\epsilon}$. However, for simplicity we shall mostly discuss the particular case of a rotation about the x_3-axis,

$$\begin{aligned} x_1' &= x_1 - \epsilon x_2 \\ x_2' &= x_2 + \epsilon x_1. \end{aligned} \tag{2.13}$$

Under such rotations, the Lagrangian $\mathcal{L}$ will be assumed to be *invariant*; from the invariance, or symmetry, a conservation law will follow – that of angular momentum. The invariance of $\mathcal{L}$ under (2.12) is expressed by the statement

$$\mathcal{L}(\mathbf{x}') = \mathcal{L}(\mathbf{x}). \tag{2.14}$$

$\mathcal{L}$ is not, however, an explicit function of the coordinates $\mathbf{x}$, but involves $\mathbf{x}$ only through $\phi(\mathbf{x})$ and $\partial_\mu\phi(\mathbf{x})$.† These functions will change under the transformations (2.12) or (2.13); their changes will depend on the Lorentz character of ϕ, or generally of all the fields present, ϕ_r. Let us confine ourselves to one *scalar* field. Then ϕ has only a single component, and we have

$$\phi'(\mathbf{x}') = \phi(\mathbf{x}). \tag{2.15}$$

(2.15) means that, under the coordinate transformation, ϕ must change to

† Indeed, from the invariance of $\mathcal{L}$ under a translation of the coordinates, another conservation law can be derived, that of energy momentum; but this would delay us, and is anyway not so useful in what follows as the rotational invariance case.

a new function ϕ', such that ϕ', evaluated at the transformed coordinate $\mathbf{x}'$, is equal to ϕ evaluated at $\mathbf{x}$. In this way, the two coordinate systems $\mathbf{x}$ and $\mathbf{x}'$ can be consistently used. Note, by contrast, that we did not put a prime on the $\mathcal{L}$ appearing on the left-hand side of (2.14), since $\mathcal{L}$ is taken to be an *invariant* function under rotations: its form does not change.

We may find the form of the new function ϕ' by using (2.15) in the form

$$\phi'(\mathbf{x}) = \phi(\mathbf{x} - \boldsymbol{\epsilon} \times \mathbf{x}) \tag{2.16}$$

$$= \phi(x_1 + \epsilon x_2, x_2 - \epsilon x_1, x_3), \tag{2.17}$$

for the special case (2.13). Expanding (2.17) by Taylor's theorem, we obtain

$$\phi'(\mathbf{x}) = \phi(\mathbf{x}) - \mathrm{i}\epsilon L^3\phi, \tag{2.18}$$

where the operator L^3 is the familiar one

$$L^3 = -\mathrm{i}x_1\partial^2 + \mathrm{i}x_2\partial^1 \tag{2.19}$$

$$= (\mathbf{x} \times \mathbf{p})^3, \tag{2.20}$$

i.e. the third component of the quantum-mechanical orbital angular momentum operator. If ϕ had several Lorentz components, an appropriate spin matrix would be added to L^3 in (2.18). (2.18) gives us the variation of ϕ under (2.13):

$$\delta\phi = -\mathrm{i}\epsilon L^3\phi. \tag{2.21}$$

We can now calculate the variation of $\mathcal{L}(\phi, \partial_\mu\phi)$ when ϕ changes by (2.21). This variation, however, is *zero*, by (2.14):

$$0 = \delta\mathcal{L} = \frac{\partial\mathcal{L}}{\partial\phi}\delta\phi + \frac{\partial\mathcal{L}}{\partial(\partial_\mu\phi)}\delta(\partial_\mu\phi). \tag{2.22}$$

Using the result $\delta(\partial_\mu\phi) = \partial_\mu(\delta\phi)$, and the Lagrange equation (2.5), we may write (2.22) as

$$0 = \epsilon\partial_\mu M^{3\mu}. \tag{2.23}$$

Since this must be true for all ϵ, we can infer that the vector quantity

$$M^{3\mu} = \frac{\partial\mathcal{L}}{\partial(\partial_\mu\phi)} \cdot -\mathrm{i}L^3\phi \tag{2.24}$$

is *divergenceless*:

$$\partial_\mu M^{3\mu} = 0. \tag{2.25}$$

Equation (2.25) has the form of a *differential conservation law*, analogous to the familiar one of electromagnetic current conservation

$$\partial_\mu j^\mu = 0. \tag{2.26}$$

Just as (2.26) implies that the charge Q, defined as the integral over all space of j^0,

$$Q = \int j^0\,\mathrm{d}^3\mathbf{x}, \tag{2.27}$$

is independent of time (i.e. is a conserved constant of motion), so also (2.25) implies that the quantity

$$M^3 = \int M^{30}\, \mathrm{d}^3\mathbf{x} \tag{2.28}$$

is conserved. In more detail, we have

$$\begin{aligned}\frac{\mathrm{d}M^3}{\mathrm{d}t} &= \int \mathrm{d}^3\mathbf{x}\ \partial_0 M^{30}(\mathbf{x}, t)\\ &= -\int \mathrm{d}^3\mathbf{x}\ \partial_i M^{3i}(\mathbf{x}, t),\end{aligned} \tag{2.29}$$

which can be transformed to a surface integral, which vanishes provided the fields fall off fast enough at infinity. Using (2.8), we may write M^3 alternatively as

$$M^3 = \int -\mathrm{i}\pi L^3 \phi\ \mathrm{d}^3\mathbf{x}, \tag{2.30}$$

which clearly generalises to

$$\mathbf{M} = \int -\mathrm{i}\pi \mathbf{L}\phi\ \mathrm{d}^3\mathbf{x}. \tag{2.31}$$

Thus with the rotational invariance of $\mathcal{L}$ is associated the conservation of the three quantities **M**, the angular momentum present in the field. This result is a special case of Noether's theorem, which states that, if $\mathcal{L}$ is invariant under a continuous one-parameter transformation, there will exist a four-vector 'current' which is differentially conserved, the spatial integral of whose zero component yields a conserved 'charge'. There are three such quantities in the present case, corresponding to the three parameters $\boldsymbol{\epsilon}$.

In addition to being conserved, the 'charges' corresponding to an invariance of $\mathcal{L}$ have a further basic property which is that they *generate* the infinitesimal transformation in question. In classical field theory, this means that the Poisson bracket of a charge and the field gives the corresponding change in the field (divided by the infinitesimal parameter). In the quantum case, the charges (which of course are operators, being functions of the fields) would be expected to be the generators of unitary transformations on the field operators, and on the Hilbert space of states. Thus in the present case of rotations, we would expect that

$$\phi' = U\phi U^{-1}, \tag{2.32}$$

where

$$U = 1 + \mathrm{i}\epsilon M^3, \tag{2.33}$$

for the rotation (2.13), or

$$U = \exp(\mathrm{i}\mathbf{a}\cdot\mathbf{M}), \tag{2.34}$$

for a finite rotation. From (2.32), (2.33) and (2.18) we obtain

$$[M^3, \phi] = -L^3\phi \tag{2.35}$$

or

$$[\mathrm{i}\epsilon M^3, \phi] = \delta\phi, \tag{2.36}$$

and M^3 is clearly the generator of the transformation $\phi \to \phi + \delta\phi$. However, M^3 has already been determined to have the form (2.30), and the commutation relation between M^3 and ϕ is also therefore determined once the canonical commutation relations are assumed. There is thus a question of consistency. In the present case (and in all cases with which we shall deal) all is well: we find

$$[M^3, \phi(\mathbf{x}', t')] = \left[\int -\mathrm{i}\pi(\mathbf{x}, t')L^3\phi(\mathbf{x}, t')\,\mathrm{d}^3\mathbf{x}, \phi(\mathbf{x}', t')\right] \tag{2.37}$$

(the time arguments in the integrals may be taken to be t', since actually, as we have seen, M^3 is independent of time)

$$= -\mathrm{i}\int [\pi(\mathbf{x}, t'), \phi(\mathbf{x}', t')]\,L^3\phi(\mathbf{x}, t')\,\mathrm{d}^3\mathbf{x}$$

$$= -\mathrm{i}\int \delta^3(\mathbf{x} - \mathbf{x}')L^3\phi(\mathbf{x}, t')\,\mathrm{d}^3\mathbf{x} \qquad \text{(using (2.9))}$$

$$= -L^3\phi(\mathbf{x}', t'),$$

as required.

We must particularly note, however, that symmetries are *not* invariably implementable by such unitary transformations. In fact, there is another possibility, in which the charge operators (generators) do not create normalisable states. The symmetry is then not unitarily implementable on the physical states. This is the situation which is called 'hidden symmetry', and it is now of equal importance in particle physics to the ordinary (unitarily implementable) case, which is called 'manifest symmetry'. The two situations are very different, and the 'hidden' one (especially as applied in gauge theories) is quite difficult to grasp intuitively. We therefore postpone further discussion of hidden symmetry until later in the book, Chapter 6. Until then, we shall be concerned only with the unitarily implementable case.

One other important caveat should be mentioned. In quantum field theory, ultraviolet divergences generally occur. These may be tamed, in theories which are renormalisable. But a divergence may sometimes *invalidate* the local conservation law corresponding to some symmetry of the Lagrangian apparent at the classical level. When this happens, we say that an 'anomaly' is present. Anomalies will be briefly discussed in Chapter 8.

We proceed now with a simple introduction to (manifest) internal symmetries.

2.3 Internal symmetries

Consider the particular Lagrangian density

$$\mathcal{L} = \tfrac{1}{2}(\partial_\mu \phi_1 \partial^\mu \phi_1 + \partial_\mu \phi_2 \partial^\mu \phi_2) - \tfrac{1}{2}m^2(\phi_1{}^2 + \phi_2{}^2), \tag{2.38}$$

where ϕ_1 and ϕ_2 are Hermitian scalar fields of the same mass m. Intuitively, we may regard ϕ_1 and ϕ_2 as 'coordinates', and would then expect $\mathcal{L}$ to be invariant under 'rotations' about the axis '3'. Indeed, we can immediately verify that $\mathcal{L}$ is invariant under

$$\left.\begin{aligned} \phi_1 \to \phi_1' &= \phi_1 \cos\theta - \phi_2 \sin\theta \\ \phi_2 \to \phi_2' &= \phi_1 \sin\theta + \phi_2 \cos\theta, \end{aligned}\right\} \tag{2.39}$$

or, for an infinitesimal transformation, under

$$\left.\begin{aligned} \phi_1 \to \phi_1' &= \phi_1 - \epsilon\phi_2 \\ \phi_2 \to \phi_2' &= \phi_2 + \epsilon\phi_1. \end{aligned}\right\} \tag{2.40}$$

Let us write (2.40) (in a notation deliberately paralleling (2.21))

$$\delta\phi_r \equiv \phi_r' - \phi_r = -\mathrm{i}\epsilon T_{rs}{}^3 \phi_s, \tag{2.41}$$

where $T_{12}{}^3 = -\mathrm{i} = -T_{21}{}^3$. Since $\mathcal{L}$ is invariant under (2.40), we have

$$0 = \delta\mathcal{L} = \frac{\partial\mathcal{L}}{\partial\phi_r}\,\delta\phi_r + \frac{\partial\mathcal{L}}{\partial(\partial^\mu\phi_r)}\,\partial^\mu(\delta\phi_r) \tag{2.42}$$

$$= \left[\partial_\mu\left(\frac{\partial\mathcal{L}}{\partial(\partial_\mu\phi_r)}\right)\right]\cdot -\mathrm{i}\epsilon T_{rs}{}^3\phi_s + \frac{\partial\mathcal{L}}{\partial(\partial_\mu\phi_r)}\cdot -\mathrm{i}\epsilon T_{rs}{}^3\partial_\mu\phi_s$$

$$= \epsilon\partial_\mu j^{3\mu}, \tag{2.43}$$

where

$$j^{3\mu} = -\mathrm{i}T_{rs}{}^3 \frac{\partial\mathcal{L}}{\partial(\partial_\mu\phi_r)}\,\phi_s = \phi_1\partial^\mu\phi_2 - \phi_2\partial^\mu\phi_1. \tag{2.44}$$

Since ϵ is arbitrary in (2.43), we deduce the differential conservation of the current $j^{3\mu}$,

$$\partial_\mu j^{3\mu} = 0. \tag{2.45}$$

The corresponding charge is

$$Q^3 = \int j^{30}\,\mathrm{d}^3\mathbf{x} = -\mathrm{i}T_{rs}{}^3 \int \pi_r\phi_s\,\mathrm{d}^3\mathbf{x} \tag{2.46}$$

$$= \int (\phi_1 \dot{\phi}_2 - \phi_2 \dot{\phi}_1)\, \mathrm{d}^3\mathbf{x}. \tag{2.47}$$

We can easily calculate the commutator

$$\begin{aligned}
[Q^3, \phi_r(x)] &= -\mathrm{i}T_{st}{}^3 \left[\int \pi_s(x')\phi_t(x')\, \mathrm{d}^3\mathbf{x}', \phi_r(x) \right] \\
&= -\mathrm{i}T_{st}{}^3 \cdot -\mathrm{i}\delta_{rs}\phi_t(x) \\
&= -T_{rt}{}^3 \phi_t(x).
\end{aligned} \tag{2.48}$$

If the transformation on ϕ_1, ϕ_2 is unitarily implementable by a unitary operator U (i.e. $\phi_1' = U\phi_1 U^{-1}$, etc.), we may write

$$U = 1 + \mathrm{i}\epsilon R, \tag{2.49}$$

where the generator R is Hermitian, for the infinitesimal case (2.40). Then (2.40) becomes

$$\left.\begin{aligned}
[R, \phi_1] &= \mathrm{i}\phi_2 \\
[R, \phi_2] &= -\mathrm{i}\phi_1,
\end{aligned}\right\} \tag{2.50}$$

or

$$[R, \phi_r] = -T_{rt}{}^3 \phi_t \tag{2.51}$$

as in (2.48), enabling us to identify the charge Q^3 as the generator of the transformation.

The equations have a neater form if we introduce the complex field

$$\phi^\dagger = \frac{1}{\sqrt{2}} (\phi_1 + \mathrm{i}\phi_2) \tag{2.52}$$

and its Hermitian conjugate

$$\phi = \frac{1}{\sqrt{2}} (\phi_1 - \mathrm{i}\phi_2), \tag{2.53}$$

in terms of which the Lagrangian is

$$\mathcal{L} = (\partial_\mu \phi)^\dagger (\partial^\mu \phi) - m^2 \phi^\dagger \phi. \tag{2.54}$$

Then (2.40) becomes

$$\phi' = (1 - \mathrm{i}\epsilon)\phi, \tag{2.55}$$

while $j^{3\mu}$ is

$$j^{3\mu} = \mathrm{i}(\phi^\dagger \partial^\mu \phi - (\partial^\mu \phi)^\dagger \phi) \tag{2.56}$$

and Q^3 is

$$Q^3 = \mathrm{i} \int (\phi^\dagger \dot{\phi} - \dot{\phi}^\dagger \phi)\, \mathrm{d}^3\mathbf{x}. \tag{2.57}$$

The commutation relations (2.48) become

$$[Q^3, \phi] = -\phi \tag{2.58}$$

$$[Q^3, \phi^\dagger] = \phi^\dagger. \tag{2.59}$$

The physical interpretation follows when we make the conventional normal mode expansion

$$\phi_r = \int \frac{\mathrm{d}^3\mathbf{k}}{[(2\pi)^3 2\omega]^{1/2}} [a_r(k)\,\mathrm{e}^{-\mathrm{i}k\cdot x} + a_r{}^\dagger(k)\,\mathrm{e}^{\mathrm{i}k\cdot x}], \tag{2.60}$$

where $\omega = (\mathbf{k}^2 + m^2)^{1/2}$. The commutation relations (2.9)-(2.11) become

$$[a_r(k), a_s{}^\dagger(k')] = \delta^3(\mathbf{k} - \mathbf{k}'), \tag{2.61}$$

$[a_r, a_s]$ and $[a_r{}^\dagger, a_s{}^\dagger]$ vanishing. The Hamiltonian is a sum of independent oscillators

$$H = \sum_r \int \mathrm{d}^3\mathbf{k}\,\omega a_r{}^\dagger(k) a_r(k) \tag{2.62}$$

after discarding the infinite zero-point contribution of the vacuum $|0\rangle$, defined by

$$a_r(k)|0\rangle = 0, \langle 0|0\rangle = 1. \tag{2.63}$$

Since $[H, a_r{}^\dagger(k)] = \omega a_r{}^\dagger(k)$, we find at once

$$H(a_r{}^\dagger(k)|0\rangle) = \omega(a_r{}^\dagger(k)|0\rangle), \tag{2.64}$$

showing that the operator $a_r{}^\dagger(k)$ creates a state of energy ω (and, more generally, momentum k) from the vacuum. We may correspondingly expand the ϕ and $\phi^\dagger$ of (2.52) and (2.53) as

$$\phi = \int \frac{\mathrm{d}^3\mathbf{k}}{[(2\pi)^3 2\omega]^{1/2}} [a_+(k)\,\mathrm{e}^{-\mathrm{i}k\cdot x} + a_-{}^\dagger(k)\,\mathrm{e}^{\mathrm{i}k\cdot x}] \tag{2.65}$$

$$\phi^\dagger = \int \frac{\mathrm{d}^3\mathbf{k}}{[(2\pi)^3 2\omega]^{1/2}} [a_+{}^\dagger(k)\,\mathrm{e}^{\mathrm{i}k\cdot x} + a_-(k)\,\mathrm{e}^{-\mathrm{i}k\cdot x}], \tag{2.66}$$

where

$$a_+{}^\dagger(k) = [a_1{}^\dagger(k) + \mathrm{i}a_2{}^\dagger(k)]/\sqrt{2} \tag{2.67}$$

$$a_-{}^\dagger(k) = [a_1{}^\dagger(k) - \mathrm{i}a_2{}^\dagger(k)]/\sqrt{2}. \tag{2.68}$$

We then find

$$[a_+(k), a_+{}^\dagger(k')] = [a_-(k), a_-{}^\dagger(k')] = \delta^3(\mathbf{k} - \mathbf{k}') \tag{2.69}$$

and all other commutators involving the $a_\pm$ vanish. In terms of these operators, the charge operator Q^3 is

$$Q^3 = \int \mathrm{d}^3\mathbf{k}[a_+{}^\dagger(k)a_+(k) - a_-{}^\dagger(k)a_-(k)]. \tag{2.70}$$

Clearly,

$$Q^3\,|0\rangle = 0 \tag{2.71}$$

and, since $[Q^3, a_+{}^\dagger(k)] = a_+{}^\dagger(k)$, we have

$$Q^3(a_+{}^\dagger(k)|0\rangle) = a_+{}^\dagger(k)|0\rangle, \tag{2.72}$$

just as in (2.64). The operator $a_+{}^\dagger(k)$ therefore creates a state with eigenvalue $+1$ for Q^3; similarly, $a_-{}^\dagger(k)$ creates a state with eigenvalue -1. The corresponding destruction operators are $a_+(k)$ and $a_-(k)$. Since $a^\dagger a$ is the number operator, Q^3 clearly counts $+1$ for each '+' particle in an arbitrary state, and -1 for each '−' particle. Furthermore, it is easily checked that

$$[Q^3, H] = 0, \tag{2.73}$$

thus confirming that Q^3 is a constant of the motion.

Of course, before interpreting the eigenvalues of Q^3 as in some way connected with *electromagnetic* charges, we must introduce an electromagnetic field. This will be done in the following chapter.

The foregoing can be easily generalised in a number of ways. For example, we may consider including a third scalar field ϕ_3, also with the same mass. This Lagrangian will be invariant under full 'three-dimensional rotations' (in the space of ϕ_1, ϕ_2 and ϕ_3):

$$\boldsymbol{\phi}' = \boldsymbol{\phi} + \boldsymbol{\epsilon} \times \boldsymbol{\phi}, \tag{2.74}$$

where we have introduced the vector notation for $\boldsymbol{\phi} = (\phi_1, \phi_2, \phi_3)$, and we now have three infinitesimal parameters $\boldsymbol{\epsilon}$. (2.74) can be written as

$$\delta\phi_r' \equiv \phi_r' - \phi_r = -\mathrm{i}\epsilon_a T_{rs}{}^a \phi_s, \tag{2.75}$$

where summation is implied over $a = 1, 2, 3$ and $s = 1, 2, 3$ and

$$T_{rs}{}^a = -\mathrm{i}\epsilon_{ars}. \tag{2.76}$$

We now have three divergenceless currents ($a = 1, 2, 3$)

$$j^{a\mu}(x) = -\mathrm{i}T_{rs}{}^a \frac{\partial \mathcal{L}}{\partial(\partial_\mu \phi_r)} \phi_s \tag{2.77}$$

$$= (\boldsymbol{\phi} \times \partial^\mu \boldsymbol{\phi})^a, \tag{2.78}$$

and three charges

$$Q^a = \int \mathrm{d}^3\mathbf{x}\, j^{a0}$$

$$= \int \mathrm{d}^3\mathbf{x}(\boldsymbol{\phi} \times \boldsymbol{\pi})^a \tag{2.79}$$

which are conserved

$$\dot{Q}^a = [Q^a, H]/\mathrm{i} = 0, \tag{2.80}$$

and generate the transformation (2.74) via (cf. (2.48) and (2.35))

$$[Q^a, \phi_r] = \mathrm{i}\epsilon_{ars}\phi_s. \tag{2.81}$$

We recognise (2.79) as the analogue for 'field space' of the angular momentum operators $\mathbf{L} = \mathbf{x} \times \mathbf{p}$ which generate rotations in ordinary three-dimensional space. Indeed, the charge operators Q^a, which generate our field rotations, obey just the same commutation relations as $\mathbf{L}$: using (2.79) again, one verifies that

$$[Q^a, Q^b] = \mathrm{i}\epsilon_{abc}Q^c. \tag{2.82}$$

This result holds if and only if the constants $T_{rs}{}^a$, in the field transformation, satisfy the same algebraic relations when they are regarded as elements of matrices T^a according to

$$(T^a)_{rs} \equiv T_{rs}{}^a. \tag{2.83}$$

Using (2.83) and (2.76) it easily follows that

$$[T^a, T^b] = \mathrm{i}\epsilon_{abc}T^c. \tag{2.84}$$

The commutation relations (2.82) are called the *algebra* of (the generators of) the group $SU(2)$. In general, the invariance of a Lagrangian under transformations associated with a unitary symmetry group allows the definition of a set of conserved charge operators which provide a field-theoretic representation of the generators of the associated Lie algebra. Symmetries of this kind, whose charges obey non-trivial commutation relations, are called *non-Abelian* symmetries. By contrast, the simple one-parameter phase transformations form the elements of the (Abelian) $U(1)$ group.

The matrices T^a form a *matrix representation* of the algebra, in that they obey the same commutation relation as the generators. In fact, they form a rather special representation, called the 'regular' or 'adjoint' representation. More generally, there will be a number of generators Q^α ($\alpha = 1, 2, \ldots, n$), which obey an algebra

$$[Q^\alpha, Q^\beta] = \mathrm{i}c_{\alpha\beta\gamma}Q^\gamma. \tag{2.85}$$

The constants $c_{\alpha\beta\gamma}$ in (2.85) are called the 'structure constants' of the group. In the regular representation, the generators are represented by matrices t^α such that

$$(t^\alpha)_{\beta\gamma} = -\mathrm{i}c_{\alpha\beta\gamma}; \tag{2.86}$$

clearly they are of dimension $n \times n$.

The fact that, in our $SU(2)$ example, we had three generators as well as three fields was just an accident of the particular illustration. Looking at (2.81) and (2.82), we see that the commutation relations of the charges with the three fields ϕ_r have exactly the same form as the commutation relations of the charges with themselves. The commutation relations of the generators with the fields prescribe how the fields transform under an infinitesimal transformation, of the sort being considered. In general, there will be some constants on the right-hand side of (2.81), expressing what linear combination of the fields is obtained on commutation with the generators. These constants, or coefficients, fully prescribe the transformation properties of the fields under the group of transformations being considered. When the constants are, as in this case, simply the structure constants of the group, we say that the *fields transform as the regular representation.*

In general, of course, many other representations will be possible. For the case of $SU(2)$, the problem of finding all of them is solved by noting that the algebra is indeed the familiar one of the angular momentum operators, so that the known results about the allowed states, and the associated matrices representing the operators, can at once be used. For example, we may consider, instead of a triplet of fields ϕ a doublet $\begin{pmatrix} u \\ d \end{pmatrix}$, which transforms by

$$\begin{pmatrix} u \\ d \end{pmatrix}' = (1 - \mathrm{i}\boldsymbol{\epsilon} \cdot \boldsymbol{\tau}/2)\begin{pmatrix} u \\ d \end{pmatrix} \tag{2.87}$$

under the infinitesimal $SU(2)$ transformations being considered. Here, the 3×3 matrices appropriate to the regular representation have been replaced by the 2×2 matrices $\boldsymbol{\tau}/2$, appropriate to the doublet case. For variety, we will consider u and d to be fermion fields (quarks), in which case a Lagrangian invariant under (2.87) would be

$$\mathcal{L} = \bar{u}(\mathrm{i}\partial\!\!\!/ - m)u + \bar{d}(\mathrm{i}\partial\!\!\!/ - m)d. \tag{2.88}$$

The associated currents are

$$j^{a\mu} = \tfrac{1}{2}\bar{\psi}\gamma^{\mu}\tau^{a}\psi, \tag{2.89}$$

where $\psi = \begin{pmatrix} u \\ d \end{pmatrix}$, and the charges are

$$Q^a = \frac{1}{2}\int \psi^{\dagger}\tau^{a}\psi \; \mathrm{d}^3\mathbf{x}. \tag{2.90}$$

We find (using now anticommutation relations for fermion fields)

$$[Q^3, u] = -\tfrac{1}{2}u \tag{2.91}$$

$$[Q^3, u^{\dagger}] = \tfrac{1}{2}u^{\dagger} \tag{2.92}$$

so that $u^{\dagger}$ creates a field with eigenvalue $+\frac{1}{2}$ of Q^3. Also, these charges Q^a (as is suggested by the notation, of course) obey the commutation relations (2.82).

Further generalisations are possible. So far we have considered only *free* fields. Assuming (see Itzykson & Zuber 1980) that a physical interpretation in terms of a vacuum, creation and annihilation operators, etc., is possible for the interacting case, we can introduce interactions which still preserve the symmetry. For example, we might consider the Lagrangian

$$\mathcal{L} = (\partial_\mu\phi)^\dagger(\partial^\mu\phi) - \tfrac{1}{2}\mu^2\phi^\dagger\phi - \tfrac{1}{2}\lambda^2(\phi^\dagger\phi)^2 \tag{2.93}$$

(one we shall have occasion to return to again), where ϕ is a *boson* field transforming as an $SU(2)$ spinor:

$$\phi = \begin{pmatrix}\phi_\text{u}\\ \phi_\text{d}\end{pmatrix} \tag{2.94}$$

with transformation

$$\delta\phi = -\mathrm{i}\epsilon\cdot\tau/2\,\phi. \tag{2.95}$$

(2.93) is certainly invariant under (2.95) and includes, in addition to the mass term $-\frac{1}{2}\mu^2\phi^\dagger\phi$, a quartic coupling with strength controlled by the parameter λ^2. In this case the Noether current is

$$j^{a\mu} = \mathrm{i}\left(\phi^\dagger\frac{\tau^a}{2}\partial^\mu\phi - (\partial^\mu\phi)^\dagger\frac{\tau^a}{2}\phi\right). \tag{2.96}$$

Once again, of course, the associated charges obey the algebra of (2.82).

As a final example (for the moment) we may mention an interaction of the type

$$\bar{\psi}\tau\psi\cdot\phi, \tag{2.97}$$

where ψ is a fermion isospinor and ϕ is a boson isovector. (2.97) is invariant under $SU(2)$ isospin transformations. Interpreting ψ as the nucleon doublet $\begin{pmatrix}p\\ n\end{pmatrix}$, and ϕ as a meson field, these considerations (extended to other hadrons, and modified to take account of the pseudoscalar nature of the pions, for example) would lead to a formalism appropriate to hadronic isospin, in which the 'charges' are just the three isospin operators **I**.

In all cases, if the symmetry is unitarily implemented in the Hilbert space of states, the charges will be well-defined operators acting in that space. As in the familiar angular momentum case, 'raising' and 'lowering' operators can be found (linear combinations of the charges), which take us from one state in a given representation to another state in the same representation. If the charges all commute with the Hamiltonian (corresponding to the currents being divergenceless and the symmetry exact), these states will all have the same energy (or mass) - i.e. we shall have degenerate multiplets. Of course, equal masses were explicitly inserted into (2.38), (2.88) and (2.93). We shall learn in Chapter 6, however, that a different possibility also exists: the Lagrangian may exhibit a symmetry, with the corresponding currents being divergenceless, but the symmetry is not unitarily implementable, and there will then be no multiplets to be seen, in general.

2.4 The Gell-Mann–Levy method

We end this chapter by introducing an alternative procedure for finding the currents corresponding to a given symmetry transformation on the fields. This procedure, first given by Gell-Mann & Levy (1960), is useful because it provides a convenient definition of the currents, and a very quick way of finding their divergences, even in cases where (as often occurs) the symmetry is not exact. It also introduces us to the idea of *global* and *local* invariance

Consider a Lagrangian $\mathcal{L}(\phi_r, \partial_\mu\phi_r)$, where we subsume fermi and bose fields under the same notation. Let the fields change by an infinitesimal *space-time-dependent* transformation

$$\phi_r(x) \to \phi_r'(x) = \phi_r(x) - \mathrm{i}\epsilon_\alpha(x)F_r{}^\alpha(\phi). \tag{2.98}$$

We emphasise again that the infinitesimal parameters $\epsilon_\alpha(x)$ in (2.98) depend on x, unlike the ϵ in (2.74) and (2.87). Transformations such as (2.74) and (2.87), which are independent of the space-time argument of the field, are called *global* transformations; those of the form (2.98), in which the parameters depend on x, are called *local* transformations (they change with the local argument of the field). Local transformations are crucially important in gauge theories, as we shall see in the following chapter. Of course, we should note that we are not considering transformations of the space-time coordinates themselves, as we did in Section 2.2.

Under the transformation (2.98), $\mathcal{L}$ changes by

$$\delta\mathcal{L} = \frac{\partial\mathcal{L}}{\partial\phi_r}(-\mathrm{i}\epsilon_\alpha(x)F_r{}^\alpha(\phi)) + \frac{\partial\mathcal{L}}{\partial(\partial_\mu\phi_r)}\partial_\mu(-\mathrm{i}\epsilon_\alpha(x)F_r{}^\alpha(\phi)) \tag{2.99}$$

$$= -\mathrm{i}\epsilon_\alpha(x)F_r{}^\alpha(\phi)\cdot\partial_\mu\left(\frac{\partial\mathcal{L}}{\partial(\partial_\mu\phi_r)}\right) + \frac{\partial\mathcal{L}}{\partial(\partial_\mu\phi_r)}\partial_\mu(-\mathrm{i}\epsilon_\alpha(x)F_r{}^\alpha(\phi)), \tag{2.100}$$

where we used the equation of motion in (2.100). (2.100) may be written as

$$\delta\mathcal{L} = j^{\alpha\mu}(x)\partial_\mu\epsilon_\alpha(x) + \epsilon_\alpha(x)\partial_\mu j^{\alpha\mu}(x), \tag{2.101}$$

where (cf. (2.44))

$$j^{\alpha\mu}(x) = -\mathrm{i}F_r{}^\alpha(\phi)\frac{\partial\mathcal{L}}{\partial(\partial_\mu\phi_r)} = \frac{\partial(\delta\mathcal{L})}{\partial(\partial_\mu\epsilon_\alpha(x))} \tag{2.102}$$

and

$$\partial_\mu j^{\alpha\mu}(x) = \frac{\partial(\delta\mathcal{L})}{\partial\epsilon_\alpha}. \tag{2.103}$$

We see at once from (2.101), that if $\mathcal{L}$ is invariant under the *global* form (constant ϵ_α) of (2.98), then the current defined by (2.102) is conserved. As an illustration of the use of (2.101), take $\mathcal{L}$ as in (2.54), with the transformation which is the local version of (2.55),

$$\phi'(x) = (1 - \mathrm{i}\epsilon(x))\phi(x). \tag{2.104}$$

We find

$$\delta \mathcal{L} = \partial_\mu \epsilon(x) \cdot \mathrm{i}(\phi^\dagger \partial^\mu \phi - (\partial^\mu \phi)^\dagger \phi), \tag{2.105}$$

and the current is read off as

$$j^{3\mu} = \mathrm{i}(\phi^\dagger \partial^\mu \phi - (\partial^\mu \phi)^\dagger \phi), \tag{2.106}$$

the same as in (2.56), with a vanishing divergence, since $[\partial(\delta\mathcal{L})]/[\partial\epsilon(x)] = 0$.

To illustrate the usefulness of the method when the symmetry is not exact, consider the Lagrangian (2.88), with *different* masses for the u and d fields:

$$\mathcal{L} = \bar{u}(\mathrm{i}\partial\!\!\!/ - m_\mathrm{u})u + \bar{d}(\mathrm{i}\partial\!\!\!/ - m_\mathrm{d})d. \tag{2.107}$$

We may write this in matrix form as

$$\mathcal{L} = \bar{\psi}(\mathrm{i}\partial\!\!\!/ - m)\psi, \tag{2.108}$$

where the mass matrix m is (in the two-component space of ψ)

$$m = \tfrac{1}{2}m_0 + \tfrac{1}{2}\delta m \tau_3, \tag{2.109}$$

with $m_0 = m_\mathrm{u} + m_\mathrm{d}$, $\delta m = m_\mathrm{u} - m_\mathrm{d}$. Under the *local* version of (2.87), namely

$$\psi'(x) = (1 - \mathrm{i}\tau \cdot \epsilon(x)/2)\psi(x), \tag{2.110}$$

we find

$$\delta\mathcal{L} = \tfrac{1}{2}\bar{\psi}\tau\gamma^\mu\psi \cdot \partial_\mu\epsilon(x) - \frac{\mathrm{i}}{2}\bar{\psi}\tau m\psi \cdot \epsilon(x) + \frac{\mathrm{i}}{2}\bar{\psi}m\tau\psi \cdot \epsilon(x), \tag{2.111}$$

and consequently we may define currents

$$j^{a\mu}(x) = \tfrac{1}{2}\bar{\psi}\gamma^\mu\tau^a\psi \tag{2.112}$$

with difergences given by

$$\partial_\mu j^{a\mu}(x) = \frac{\mathrm{i}}{2}\bar{\psi}[m, \tau]\psi. \tag{2.113}$$

Clearly, the third component, $a = 3$, is conserved in this case (τ_3 commutes with m of (2.109)), but $\partial_\mu j^{1\mu}(x)$ and $\partial_\mu j^{2\mu}(x)$ are both proportional to the mass difference δm, and so these currents are only conserved (as we would expect) if $\delta m = 0$.

Note, however, that even if $j^{1\mu}$ and $j^{2\mu}$ are not conserved, the currents are still defined by (2.112); in particular, the associated charges will still obey the algebraic commutation relations (2.82). The fact that the charges and (with some possible reservations) the current operators obey the same commutation relations, even if the symmetry is not exact, suggests that these commutation relations might be a sensible thing to focus attention on in those physical cases where the symmetry appears to be only approximate: the *algebra* could then still be exact. This is the starting point of the idea of current algebra, due to Gell-Mann (see further Section 6.8).

3 Gauge fields and the gauge principle

In this chapter we shall show how the requirement of invariance of the matter Lagrangian under *local phase transformations* leads to the introduction of vector fields (*gauge fields*) interacting with the matter fields in a definite way. In the non-Abelian case, the self interactions of the vector fields are also prescribed.

3.1 The inclusion of electromagnetism: gauge-invariance

The conventional way of introducing the electromagnetic interaction in quantum mechanics, or quantum field theory, is via the so-called 'minimal prescription', whereby the momentum operator p^μ is replaced by $p^\mu - qA^\mu$, for a particle of charge q, where A^μ is the vector potential. The corresponding classical Hamiltonian then reproduces, via Hamilton's equations, the correct Lorentz force; in quantum mechanics, since p^μ is replaced by $\mathrm{i}\partial^\mu$, the prescription is

$$\partial^\mu \to \partial^\mu + \mathrm{i}qA^\mu. \tag{3.1}$$

The combination $\partial^\mu + \mathrm{i}qA^\mu$ is of fundamental importance, and is called the 'covariant derivative', denoted by D^μ:

$$D^\mu \equiv \partial^\mu + \mathrm{i}qA^\mu. \tag{3.2}$$

The significance of D^μ will emerge as we proceed.

The rule $\partial^\mu \to D^\mu$ may be taken over to quantum field theory. For example, the Lagrangian for a free Dirac particle of mass m is

$$\mathcal{L}_0 = \bar{\psi}(\mathrm{i}\partial\!\!\!/ - m)\psi, \tag{3.3}$$

which becomes

$$\mathcal{L}_1 = \mathcal{L}_0 + \mathcal{L}_{\text{int}} = \bar{\psi}(\mathrm{i}\partial\!\!\!/ - m)\psi - q\bar{\psi}\gamma^\mu\psi A_\mu \tag{3.4}$$

after the replacement (3.1), if the field ψ corresponds to particles of charge q. To obtain the complete Lagrangian, we must of course add to (3.4) a part which yields the Maxwell equations for the potentials A^μ; we shall defer more detailed consideration of this until the following chapter, since our main concern in this chapter is with symmetries, merely noting here that the standard classical Lagrangian for the electromagnetic field would be

$$\mathcal{L}_{em} = -\tfrac{1}{4}F_{\mu\nu}F^{\mu\nu}, \tag{3.5}$$

where

$$F^{\mu\nu} = \partial^\mu A^\nu - \partial^\nu A^\mu. \tag{3.6}$$

The Euler–Lagrange equations, applied to the complete Lagrangian $\mathcal{L} = \mathcal{L}_{em} + \mathcal{L}_0 + \mathcal{L}_{int}$ then yield the field equations for A^μ

$$\Box A^\mu - \partial^\mu(\partial_\nu A^\nu) = j^\mu, \tag{3.7}$$

where j^μ is the current $q\bar{\psi}\gamma^\mu\psi$; (3.7) certainly has the correct Maxwell form.

The electromagnetic field has been introduced here via the four-vector potential A^μ, as is usual. The physical electric and magnetic fields are related to $A^\mu = (V, \mathbf{A})$ by

$$\mathbf{E} = -\nabla V - \frac{\partial \mathbf{A}}{\partial t} \tag{3.8}$$

and

$$\mathbf{B} = \nabla \times \mathbf{A}. \tag{3.9}$$

V and $\mathbf{A}$ are not uniquely defined by (3.8) and (3.9), however: indeed, a transformation of the form

$$A^\mu \to A'^\mu = A^\mu + \partial^\mu\chi \tag{3.10}$$

can be made, for any differentiable scalar function, without altering $\mathbf{E}$ and $\mathbf{B}$. This is the property which is called the gauge-invariance of the Maxwell equations: (3.10) is a gauge transformation on the vector potential.

Since (3.10) is an invariance of the classical Maxwell equations, we must expect it to be an invariance also in the quantum theory. Yet, at first sight, this seems not to be the case. Certainly, the Lagrangian for the electromagnetic field alone (3.5) is invariant under (3.10) - $F_{\mu\nu}$ being a kind of four-dimensional 'curl', which is unchanged by the addition of a gradient. However, the part $\mathcal{L}_{int}$ certainly changes under (3.10). If it changed merely by a total derivative this would not change the field equation derived from the action principle which involves only the integral of $\mathcal{L}$. However, $\delta\mathcal{L}_{int} = -q\bar{\psi}\gamma^\mu\psi\partial_\mu\chi$ is not a total derivative. The only possibility is that this change in $\mathcal{L}_{int}$ must be cancelled by a corresponding change in the free part, $\mathcal{L}_0$.

This cancellation can be achieved if it is supposed that, when A^μ undergoes the gauge transformation (3.10), the field ψ also transforms, by a *local phase transformation* of the specific form

$$\psi(x) \to \psi'(x) = \exp(-iq\chi(x))\psi(x). \tag{3.11}$$

It is easily verified that, under (3.11), $\mathcal{L}_0$ of (3.3) changes by

$$\delta\mathcal{L}_0 = q\bar{\psi}\gamma^\mu\psi\partial_\mu\chi, \tag{3.12}$$

thereby precisely cancelling $\delta\mathcal{L}_{\text{int}}$. The full Lagrangian $\mathcal{L}_1 + \mathcal{L}_{\text{em}}$ is invariant under the combined transformations (3.10) and (3.11); these latter, then, express gauge-invariance in quantum field theory (with (3.11) generalised to any particle of charge q).

We end this section with a remark about the form of the current operators in the presence of the electromagnetic interaction. For a fermion Lagrangian of the type (3.3), the Gell-Mann–Levy procedure gives, from (3.12),

$$j^\mu = q\bar{\psi}\gamma^\mu\psi. \tag{3.13}$$

This is, of course, just the same as the current associated with a one-parameter global $U(1)$ phase invariance, in the absence of A^μ. However, the situation is more interesting in the boson case. We may apply the prescription (3.1) to the Lagrangian for the free complex scalar field

$$\mathcal{L}_0 = (\partial_\mu\phi)^\dagger(\partial^\mu\phi) - m^2\phi^\dagger\phi, \tag{3.14}$$

obtaining the interaction piece

$$\mathcal{L}_{\text{int}} = -\mathrm{i}qA_\mu[\phi^\dagger\partial^\mu\phi - (\partial^\mu\phi)^\dagger\phi] + q^2A^2\phi^\dagger\phi. \tag{3.15}$$

The Gell-Mann–Levy transformation

$$\phi(x) \to \phi'(x) = \exp(-\mathrm{i}\alpha(x))\phi(x), \tag{3.16}$$

applied to $\mathcal{L}_0 + \mathcal{L}_{\text{int}}$ of (3.14) and (3.15), then yields the current

$$j^\mu(x) = \mathrm{i}q[\phi^\dagger\partial^\mu\phi - (\partial^\mu\phi)^\dagger\phi] - 2q^2A^\mu\phi^\dagger\phi. \tag{3.17}$$

(3.17) can be written as

$$j^\mu(x) = \mathrm{i}q\{\phi^\dagger(\partial^\mu + \mathrm{i}qA^\mu)\phi - [(\partial^\mu + \mathrm{i}qA^\mu)\phi]^\dagger\phi\}, \tag{3.18}$$

which is simply the global form (2.56) made *gauge-invariant* by the covariant derivative replacement (3.1). The extra term in (3.17), not found in (3.13), occurs because of the gradient in the boson current.

It is of critical significance that, in both the cases (3.13) and (3.17), the current is actually also just

$$-\frac{\partial(\mathcal{L}_{\text{total}})}{\partial A_\mu}. \tag{3.19}$$

Hence (from the Euler–Lagrange equations), it is the quantity appearing on the 'right-hand side' of the Maxwell equations for A^μ. This is what allows us, finally, to interpret (3.13) and (3.17) as electromagnetic currents. Gauge field theory is unique in that the currents play a *dual* role; they are both the symmetry currents in the Noether sense, and also the 'sources' of the vector fields.

3.2 Global and local phase invariance: the gauge principle

We now turn the preceding discussion upside down and introduce the idea which is crucial to the development of gauge theories.

As we have seen in the previous chapter, global phase invariances are common in particle physics, being associated with various symmetries, and hence with conservation laws via Noether's theorem. The possibility of changing the phase of a field arbitrarily means that its phase is a matter of convention. In the case of the $SU(2)$ transformations concerned in Section 2.3, a geometrical interpretation of them is often made - indeed, we used it ourselves in talking about 'rotations' in the space of the fields. To be definite, let us now call this the $SU(2)$ of isospin. Invariance under global $SU(2)$ transformations of the type (2.74) or (2.87) is closely analogous to a kind of 'rotation invariance'. One way of putting the physical consequence of such invariance is that we may choose the 'axes in isospace' as we please. In other words, the definition of the fields to be associated with the proton, say, ($I_3 = \frac{1}{2}$) and with the neutron ($I_3 = -\frac{1}{2}$) is, in the limit of exact $SU(2)$-invariance, entirely conventional up to a unitary transformation.† However, a *global* symmetry of this type implies that once we have decided what the convention ought to be at one space-time point, the same convention must be followed at all other space-time points, because the transformation is not allowed to vary from point to point. Yang & Mills (1954) were the first to question whether this was entirely reasonable. 'It seems', they wrote, 'that this is not consistent with the localized field concept that underlies the usual physical theories. In the present paper we wish to explore the possibility of requiring all interactions to be invariant under *independent* rotations of the isotopic spin at all space-time points.' Thus was the study of the remarkable non-Abelian gauge field theories (see Section 3.3) initiated.

The proposal of Yang & Mills amounts to the requirement that the theory be invariant under *local* phase transformations. Let us see how this would operate in the very simple case of a $U(1)$ transformation involving a single phase parameter. We start with the Lagrangian

$$\mathcal{L}_0 = \bar{\psi}(\mathrm{i}\not{\partial} - m)\psi, \tag{3.20}$$

say, which is certainly invariant under the global phase transformation

$$\psi(x) \to \psi'(x) = \exp(-\mathrm{i}\alpha)\psi(x). \tag{3.21}$$

Let us 'explore the possibility' of invariance under the local phase transformation

$$\psi(x) \to \psi'(x) = \exp(-\mathrm{i}\alpha(x))\psi(x). \tag{3.22}$$

This step is formally the same as in the Gell-Mann–Levy procedure, but now we

† Note however, that the electromagnetic interaction is *not* $SU(2)$-invariant: hence it distinguishes between p and n.

are going to demand that a full Lagrangian $\mathcal{L}_1$ exists which is actually *invariant* under (3.22). Clearly, $\mathcal{L}_0$ changes by

$$\delta\mathcal{L}_0 = \bar{\psi}\gamma^\mu\psi\,\partial_\mu\alpha(x). \tag{3.23}$$

If we now require that $\delta\mathcal{L}_1 = 0$, then $\mathcal{L}_1$ must contain a term in addition to $\mathcal{L}_0$, whose change under (3.22) exactly cancels $\delta\mathcal{L}_0$. Such an $\mathcal{L}_1$ is, of course,

$$\mathcal{L}_1 = \mathcal{L}_0 - q\bar{\psi}\gamma^\mu\psi A_\mu, \tag{3.24}$$

provided that, when ψ undergoes (3.22), A_μ changes by

$$A_\mu(x) \to A'_\mu(x) = A_\mu(x) + \frac{1}{q}\partial_\mu\alpha(x). \tag{3.25}$$

Writing $\alpha(x)$ as $q\chi(x)$ we recover precisely (3.10) and (3.11), and the original 'minimal prescription' Lagrangian. The introduction of a *vector* field, transforming according to (3.25), would seem to be necessary if we are to have local freedom as to phase convention. Furthermore, the vector field has to be one such that transformations of the form (3.25) can be performed on it without altering the physical results (since (3.25) is part of the local *invariance* of $\mathcal{L}_1$); such vector fields are called *gauge fields.* We may say that, at least in this simple case, the requirement of local phase invariance has 'generated' the interaction term $-q\bar{\psi}\gamma^\mu\psi A_\mu$ between the matter field ψ and the gauge field A_μ. This is the essence of the gauge principle for generating dynamical theories: suitable gauge fields are introduced, with interactions such that the required local invariance holds.

We must note at once, however, that if the (gauge) transformation (3.25) on A_μ is indeed to be an invariance of the Lagrangian - when combined with (3.22) - we can *not*, apparently, allow the vector field A_μ to have a mass. This would enter $\mathcal{L}$ in the form $\frac{1}{2}m^2A^2$ (see Chapter 5 below for further comment about massive vector fields), which is quite clearly not going to be invariant under (3.25). Hence, it would appear that this remarkable idea for generating interactions is restricted to massless vector fields only.

The original paper of Yang & Mills (1954) generalised invariance under the local one-parameter type of phase transformation (3.22) to invariance under local isospin-type transformations - that is, under transformations of the type

$$\psi \to \psi' = \exp\left(-\mathrm{i}g\theta_a(x)t^a\right)\psi(x), \tag{3.26}$$

where the matrices t^a $(a = 1, 2, 3)$ represent the algebra of $SU(2)$ in the representation appropriate to the multiplet ψ. Such invariance will now demand the introduction of three gauge fields $A_\mu{}^a(x)$, which will transform in some way analogous to (3.25), when ψ undergoes the local phase transformation (3.26). The actual form is somewhat more complicated than (3.25), and will be given in Section 3.3, where we shall also discuss how the local invariance determines the interactions in this case.

Here too, however, it would seem that the vector fields $A_\mu{}^a(x)$ have to be massless. Yang & Mills themselves introduced their idea in the context of hadronic isospin. They hoped to explain hadronic interactions as being a consequence of the local version of isospin invariance (whose global consequences were manifest, at least approximately, in the particle multiplets). But this seemed a rather unlikely prospect at the time, even before the eventual discovery of strongly interacting vector particles (ρ, ω, ϕ, . . .), because it was always clear that the hadronic force was short ranged, and could not be mediated by massless gauge quanta. Nevertheless, the idea was boldly pursued by Sakurai (1960).

From a vantage point somewhat later than 1954, one could have argued, perhaps, that the gauge principle of generating strong interactions should apply at a more fundamental level than that of the (clearly composite) hadrons - rather, to the interactions between their constituents (as is now believed to be the case - QCD is precisely such a gauge theory of quark interactions: see Section 3.4). However, the weak interactions of *leptons* continued to appear to be fundamental, but hopes of describing them in terms of gauge field theory seemed, until 1964, doomed to disappointment for the same reason as in hadron physics: the mediating vector quanta had to be massive, as we saw in Section 1.1. The breakthrough came with the realisation that the gauge principle could still be at work, but the associated symmetry could be 'hidden' (or 'spontaneously broken') - i.e. not unitarily implemented. It turned out that, in this case, the gauge fields could acquire a mass, as we shall discuss in detail later. Furthermore, it was then proved ('t Hooft 1971*b*) that the gauge-invariance of such theories, though hidden, was still sufficient to prove renormalisability, and thus, for the first time, a calculable theory of weak interactions became possible. It now seems very likely indeed that the Glashow-Salam-Weinberg theory of weak interactions (or some generalisation of it), which employs a Yang-Mills type of gauge field theory (utilising a 'weak isospin' symmetry† - see Chapter 7) in the hidden symmetry realisation, describes correctly weak interaction phenomena.

We postpone discussion of the hidden symmetry case until Chapter 6; we end the present chapter with a first look at non-Abelian (Yang-Mills) gauge fields.

3.3 Local non-Abelian symmetry

We proceed to investigate the possibility of Lagrangians invariant under local phase transformations of the type (3.26), which generalise the global non-Abelian symmetries (such as $SU(2)$) considered in Section 2.3.

Rather than present the general case, we follow our informal practice of treating an illustrative model first. We consider again the Lagrangian (introduced

† *Not* the familiar hadronic isospin symmetry.

in Section 2.3)

$$\mathcal{L} = (\partial_\mu \phi)^\dagger (\partial^\mu \phi) - \tfrac{1}{2}\mu^2 \phi^\dagger \phi - \tfrac{1}{2}\lambda^2 (\phi^\dagger \phi)^2 \tag{3.27}$$

describing a boson isospinor ϕ. As we saw in Section 2.3, (3.27) is invariant under the global isospin transformation

$$\delta\phi(x) = -\mathrm{i}\epsilon \cdot \tau/2\, \phi(x) \tag{3.28}$$

with constant ϵ. We seek to make it invariant under the local version of (3.28), namely,

$$\delta\phi(x) = -\mathrm{i}g\epsilon(x) \cdot \tau/2\, \phi(x), \tag{3.29}$$

where, in conformity with (3.26), we have now slipped in a constant g which, by analogy with the electromagnetic case, will be interpreted as a coupling strength. We choose this particular model, incidentally, because it will turn out to be an essential part of the GSW theory.

Clearly, (3.27) is *not* invariant under (3.29) because of the $\partial_\mu \epsilon(x)$ terms which arise exactly as in (3.23). Equally clearly, we expect that, by introducing gauge fields with suitable transformation properties, we can cancel these terms. We now need three gauge fields, let us call them $W^{a\mu}(x)$, since there are three gradient terms $\partial^\mu \epsilon^a(x)$. The required prescription is actually a simple generalisation of the 'minimal replacement' - or covariant derivative replacement - recipe we used in the one-parameter case of electromagnetism.

The crucial property of the covariant derivative in the electromagnetic ($U(1)$) case,

$$D^\mu = \partial^\mu + \mathrm{i}qA^\mu, \tag{3.30}$$

is that the quantity $D^\mu \psi(x)$ transforms under the combined *local* transformations

$$\left.\begin{aligned} \psi(x) &\to \psi'(x) = \exp(-\mathrm{i}q\chi(x))\psi(x) \\ A^\mu(x) &\to A'^\mu(x) = A^\mu(x) + \partial^\mu \chi(x), \end{aligned}\right\} \tag{3.31}$$

in exactly the same way as $\partial^\mu \psi(x)$ does under the *global* transformation:

$$\psi(x) \to \psi'(x) = \exp(-\mathrm{i}q\chi)\psi(x). \qquad \text{(Constant } \chi.) \tag{3.32}$$

This property is easily checked:

$$\begin{aligned} (D^\mu \psi(x))' &= (\partial^\mu + \mathrm{i}qA'^\mu(x))(\exp(-\mathrm{i}q\chi(x))\psi(x)) \\ &= -\mathrm{i}q\,\partial^\mu \chi(x) \cdot (\exp(-\mathrm{i}q\chi(x))\psi(x)) + \exp(-\mathrm{i}q\chi(x))\partial^\mu \psi(x) \\ &\quad + \mathrm{i}qA^\mu(x)\exp(-\mathrm{i}q\chi(x))\psi(x) \\ &\quad + \mathrm{i}q\,\partial^\mu \chi(x) \cdot \exp(-\mathrm{i}q\chi(x))\psi(x) \\ &= \exp(-\mathrm{i}q\chi(x))(\partial^\mu + \mathrm{i}qA^\mu(x))\psi(x) \\ &= \exp(-\mathrm{i}q\chi(x))(D^\mu \psi(x)). \end{aligned} \tag{3.33}$$

It follows that any $\mathcal{L}_0$ which is invariant under the global transformation (3.32) can be made locally invariant by merely replacing ∂^μ by D^μ.

In our specific $SU(2)$ model, we therefore require some appropriate generalisation of D^μ, in order to make $\mathcal{L}$ locally $SU(2)$-invariant. The answer is that the covariant derivative for this case is given by

$$D^\mu = \partial^\mu + ig\boldsymbol{\tau}\cdot \boldsymbol{W}^\mu(x)/2. \tag{3.34}$$

This is the form appropriate for an $SU(2)$ doublet, which is the reason for the appearance of the characteristic matrices $\boldsymbol{\tau}/2$: we emphasise that D^μ in (3.34) is a *matrix* acting on the components of ϕ. The appropriate definition of D^μ for fields transforming according to other $SU(2)$ representations will be given later. The property required of (3.34) is that $D^\mu\phi$ – with ϕ an isodoublet – should transform under the local $SU(2)$ transformation (3.29) exactly as $\partial^\mu\phi$ does under the global one (2.95). This requirement determines the transformation law of the gauge fields $W^{a\mu}(x)$. Namely, we require

$$\delta(D^\mu\phi) = -ig\boldsymbol{\epsilon}(x)\cdot\boldsymbol{\tau}/2(D^\mu\phi). \tag{3.35}$$

The left-hand side is $\delta[(\partial^\mu + ig\boldsymbol{\tau}\cdot\mathbf{W}^\mu/2)\phi]$, and will involve $\delta\mathbf{W}^\mu$. It is a useful exercise to check that the (infinitesimal) transformation law of the $\mathbf{W}^\mu$ must be

$$\delta\mathbf{W}^\mu(x) = \partial^\mu\boldsymbol{\epsilon}(x) + g\boldsymbol{\epsilon}(x)\times\mathbf{W}^\mu(x). \tag{3.36}$$

The first term in (3.36) is the usual inhomogeneous part which, as in the electromagnetic gauge transformation on $A^\mu(x)$, will cancel the $\partial^\mu\boldsymbol{\epsilon}(x)$ terms in $\delta\mathcal{L}$. The second term is easy to interpret by referring to (2.74): it is merely the statement that the fields $\mathbf{W}^\mu$ are the components of an *isovector* under the local $SU(2)$.

For completeness, we shall also give the transformation law for $\mathbf{W}^\mu$ appropriate to the finite (not infinitesimal) case. We write the transformation of ϕ as

$$\phi'(x) = U(x)\phi(x), \tag{3.37}$$

with

$$U(x) = \exp(-i\boldsymbol{\alpha}(x)\cdot\boldsymbol{\tau}/2), \tag{3.38}$$

where

$$UU^\dagger = U^\dagger U = I. \tag{3.39}$$

We require

$$D'^\mu\phi'(x) = U(x)(D^\mu\phi(x)) \tag{3.40}$$

or, equivalently,

$$D'^\mu = U(x)D^\mu U^\dagger(x). \tag{3.41}$$

But D'^μ is also $\partial^\mu + i\boldsymbol{\tau}\cdot\mathbf{W}'^\mu(x)/2$. Hence

$$\partial^\mu + i\boldsymbol{\tau}\cdot\mathbf{W}'^\mu(x)/2 = U(x)[\partial^\mu + i\boldsymbol{\tau}\cdot\mathbf{W}^\mu(x)/2]U^\dagger(x). \tag{3.42}$$

If we now introduce the (dangerously compact!) notation

$$W^\mu(x) \equiv \boldsymbol{\tau}\cdot\mathbf{W}^\mu(x)/2, \tag{3.43}$$

(3.42) yields

$$W'^{\mu}(x) = -\mathrm{i}U(x)[\partial^{\mu}U^{\dagger}(x)] + U(x)W^{\mu}(x)U^{\dagger}(x). \tag{3.44}$$

It is a good exercise to recover the infinitesimal form (3.36) from (3.44).

We can now assert that any function of ϕ and $D^{\mu}\phi$ which is globally $SU(2)$-invariant is also locally $SU(2)$-invariant. There remains the question of the Lagrangian describing the fields $\mathbf{W}^{\mu}$ themselves. For the electromagnetic case we used $F^{\mu\nu} = \partial^{\mu}A^{\nu} - \partial^{\nu}A^{\mu}$, and this did *not* have an inhomogeneous term in its transformation law. This suggests that we look for an $\mathbf{F}^{\mu\nu}$ which transforms simply as an isovector under the local $SU(2)$ transformation. The answer is that if we form

$$\mathbf{F}^{\mu\nu} = \partial^{\mu}\mathbf{W}^{\nu} - \partial^{\nu}\mathbf{W}^{\mu} - g\mathbf{W}^{\mu} \times \mathbf{W}^{\nu}, \tag{3.45}$$

then

$$\delta\mathbf{F}^{\mu\nu} = g\epsilon(x) \times \mathbf{F}^{\mu\nu}, \tag{3.46}$$

under (3.36). The locally $SU(2)$-invariant $\mathbf{W}^{\mu}$ part of the Lagrangian is then

$$\mathcal{L}_W = -\tfrac{1}{4}\mathbf{F}^{\mu\nu} \cdot \mathbf{F}_{\mu\nu}, \tag{3.47}$$

the dot product being in $SU(2)$ space. The full locally $SU(2)$-invariant version of (3.27) is then

$$\mathcal{L} = (D_{\mu}\phi)^{\dagger}(D^{\mu}\phi) - \tfrac{1}{2}\mu^2\phi^{\dagger}\phi - \tfrac{1}{2}\lambda^2(\phi^{\dagger}\phi)^2 - \tfrac{1}{4}\mathbf{F}^{\mu\nu} \cdot \mathbf{F}_{\mu\nu}\,. \tag{3.48}$$

There are now a number of comments to be made about the above procedure and results.

1. In the first place, the 'covariant derivative' prescription has specified the interaction between the gauge fields $W^{a\mu}$ and the matter fields ϕ, as in the $U(1)$ case.

2. But, secondly, even if there were no matter fields at all, the 'pure Yang–Mills' part $\mathcal{L}_W$ contains *interactions.* This is not true of the electromagnetic case, where $\mathcal{L}_{\text{em}} = -\frac{1}{4}F^{\mu\nu}F_{\mu\nu}$ describes the free Maxwell theory. In fact, the notation in (3.47) is very condensed, and it is worth unpacking it into

$$\begin{aligned} -\tfrac{1}{4}\mathbf{F}^{\mu\nu} \cdot \mathbf{F}_{\mu\nu} = & -\tfrac{1}{2}(\partial_{\mu}\mathbf{W}_{\nu} - \partial_{\nu}\mathbf{W}_{\mu}) \cdot \partial^{\mu}\mathbf{W}^{\nu} \\ & + g(\mathbf{W}_{\mu} \times \mathbf{W}_{\nu}) \cdot \partial^{\mu}\mathbf{W}^{\nu} \\ & - \tfrac{1}{4}g^2\,[(\mathbf{W}_{\mu} \cdot \mathbf{W}^{\mu})^2 - (\mathbf{W}_{\mu} \cdot \mathbf{W}_{\nu})(\mathbf{W}^{\mu} \cdot \mathbf{W}^{\nu})]. \end{aligned} \tag{3.49}$$

Clearly, (3.49) involves trilinear and quadrilinear couplings, which lead to three-point vertices of the form of Fig. 3.1, and to four-point vertices (Fig. 3.2), when we follow the normal route to the Feynman rules from the Lagrangian. Thus a Yang–Mills (non-Abelian) gauge field theory is a non-linear theory, because the gauge fields enjoy self interactions – they are the 'quanta of the isospin field', and carry isospin themselves. This is a fundamental and profound difference with respect to the electromagnetic case, in which the γ quanta are *un*charged.

3. In general, the non-Abelian gauge fields must transform according to the regular representation of the local group G in question (Glashow & Gell-Mann 1961).

4. The general definition of the covariant derivative is as follows. Suppose a field ϕ transforms under G according to some p-dimensional representation

$$\delta\phi = -ig\epsilon_\alpha(x)t^\alpha\phi, \tag{3.50}$$

where the t^α ($\alpha = 1, 2, \ldots$ number of generators of G) are $p \times p$ matrices obeying the algebra of G:

$$[t^\alpha, t^\beta] = ic_{\alpha\beta\gamma}t^\gamma, \tag{3.51}$$

with structure constants $c_{\alpha\beta\gamma}$. Then

$$D^\mu\phi = \partial^\mu\phi + igt^\alpha W^{\alpha\mu}\phi \quad \text{(sum on } \alpha\text{)}. \tag{3.52}$$

The gauge fields $W^{\alpha\mu}(x)$ transform by the addition of the inhomogeneous part to (3.50), namely, by

$$\delta W^{\alpha\mu}(x) = \partial^\mu\epsilon^\alpha(x) + gc_{\alpha\beta\gamma}\epsilon^\beta(x)W^{\gamma\mu}(x), \tag{3.53}$$

since the matrices in the regular representation are just (cf. (2.86))

$$(t^\alpha)_{\beta\gamma} = -ic_{\alpha\beta\gamma}. \tag{3.54}$$

Remarkably enough, the change in $W^{\alpha\mu}(x)$ given by (3.53) is precisely the covariant derivative of $\epsilon^\alpha(x)$:

$$\delta W^{\alpha\mu}(x) = (D^\mu\boldsymbol{\epsilon}(x))^\alpha. \tag{3.55}$$

The generalisation of (3.45) is

$$F^{\alpha\mu\nu} = \partial^\mu W^{\alpha\nu} - \partial^\nu W^{\alpha\mu} - gc_{\alpha\beta\gamma}W^{\beta\mu}W^{\gamma\nu}. \tag{3.56}$$

Fig. 3.1. Three-point coupling among the gauge fields.

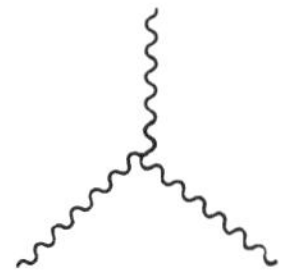

Fig. 3.2. Four-point gauge field coupling.

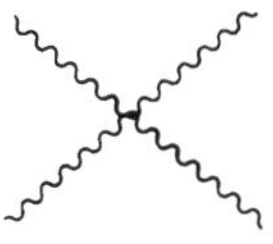

The finite transformation (3.44) was derived in terms of the matrix W^μ of (3.43), which involved the $\boldsymbol{\tau}$ matrices. Their appearance was due simply to the fact that we were considering there, as our initial example, an isospinor matter field ϕ. More generally, we can introduce a matrix

$$W^\mu = t^\alpha W^{\alpha\mu}, \tag{3.57}$$

where the t^α are the matrices representing the generators in an arbitrary representation. Equation (3.43) is then still valid for the finite gauge transformation on W^μ of (3.57). We may also define the analogous matrix

$$F^{\mu\nu} = t^\alpha F^{\alpha\mu\nu}, \tag{3.58}$$

in terms of which (3.56) may be written

$$F^{\mu\nu} = \partial^\mu W^\nu - \partial^\nu W^\mu + \mathrm{i}g[W^\mu, W^\nu], \tag{3.59}$$

in which we have used (3.51). This matrix notation may also be used to give an alternative expression for the covariant derivative, when acting on quantities R^α transforming as the regular representation of the group. We form

$$R = t^\alpha R^\alpha$$

and then (cf. (3.53) and (3.51))

$$D^\mu R = \partial^\mu R + \mathrm{i}g[W^\mu, R]. \tag{3.60}$$

5. A crucial point is that it was not through an oversight that we used only *one* 'g' in all the above equations. The whole thing will not work at all if the g's in the transformation laws for the different fields are *not* the same. This again is quite different from the $U(1)$ case. There, each charged field can, and does, couple to A_μ with its 'own' charge (e, $2e$, $-3e$, etc.). It is a useful exercise to check that, once $\delta \mathbf{W}_\mu$ has been specified with a given g by (3.53), if (3.29) and (3.34) are used with g replaced by g', (3.35) will only be obtained if $g = g'$. This exercise will reveal that the result hinges on the non-trivial *commutation* relation between $\boldsymbol{\tau}\cdot\boldsymbol{\epsilon}$ and $\boldsymbol{\tau}\cdot\mathbf{W}_\mu$, which is characteristic of a non-Abelian theory. Qualitatively, one can say that 'we can't arbitrarily change the scale of operators that have definite commutation relations'. There is therefore exactly *one* coupling constant for a Yang–Mills field transforming under a group G: we say that the gauge field couples *universally* to all matter fields.

6. Universality (in the sense of a single coupling parameter g) holds provided that the generators cannot be separated into two or more sets such that the generators in one set *commute* with those in another, but do not commute among themselves. In the case in which this does happen, we write $G = G_1 \times G_2, \ldots$ where the generators of G_1 commute with those of G_2, etc. For example, the model we have been considering, (3.48), actually has a global $U(1)$ - invariance

$$\phi \to \phi' = \exp(-\mathrm{i}\alpha)\phi \tag{3.61}$$

as well as the global $SU(2)$ one, simply because ϕ always appears multiplied by $\phi^\dagger$. The full global invariance group of (3.53) is therefore $SU(2)\times U(1)$ (and, of course, this was deliberate, since this is the invariance group of the GSW theory, as we shall see). When this symmetry is 'gauged' - as it is in GSW theory - we will be dealing with two independent coupling constants.

7. Actually, the case of $U(1)$ is somewhat special, and there is no real reason why every individual charged particle should not couple to A^μ with its own personal charge (which, as far as the $U(1)$ symmetry is concerned, need not be quantised in units of e). This is one motivation (Georgi & Glashow 1974, footnote 10) for seeking to embed a group of the type $SU(2)\times U(1)$, where $U(1)$ refers to (or involves) the electromagnetic gauge group, in a larger group which does not 'factorise' in this way: see Chapter 7 below.

8. It is interesting to calculate the Noether (Gell-Mann-Levy) current corresponding to the $SU(2)$-invariance of (3.48); the result is

$$j^{a\mu} = \mathrm{i}g\left(\phi^\dagger \frac{\tau^a}{2}\partial^\mu\phi - (\partial^\mu\phi)^\dagger\frac{\tau^a}{2}\phi\right) - \frac{g^2}{2}\phi^\dagger\phi W^{a\mu} + g(\mathbf{F}^{\mu\nu}\times\mathbf{W}_\nu)^a. \quad (3.62)$$

The first term is the ordinary global current (2.96). The second term is analogous to the second term of (3.17); as in that case, these first two terms in expression (3.62) can be obtained from the global current (2.96) by making the $SU(2)$-gauge-invariant replacement $\partial^\mu \to D^\mu$, given by (3.34). The third term (which is not gauge-invariant) corresponds to the fact that the **W**'s themselves carry this $SU(2)$ 'isospin', and therefore they would be expected to contribute to the current. As in the $U(1)$ examples discussed in Section 3.1, the current (3.62) is precisely $(-\partial\mathcal{L})/(\partial W_\mu{}^a)$, with $\mathcal{L}$ given by (3.48), and thus it provides the source of the W fields.

We now consider the application of these ideas to a case of practical interest.

3.4 A Yang-Mills theory of strong interactions: introduction to QCD

The original intention of Yang & Mills (1954), in developing non-Abelian gauge theory, was to apply it to the strong interactions between hadrons. When they made their suggestion, as we have already remarked, the chief theoretical obstacle to its acceptance was the *mass* problem: how could a theory remain gauge-invariant if the vector particles of the (strong) gauge field had *mass*? We now know that hidden gauge symmetry can provide a way out of this dilemma (see Chapter 6). But it is still true that these ideas seem not to be directly applicable to hadrons, for the following reason. In general, the gauge fields must be coupled to all the components of the current associated with the global symmetry group. If this symmetry is of the conventional, 'manifest', type (see Chapter 6 for the 'hidden' case), the gauge quanta *are* massless. So consider trying to explain the (empirically broken) $SU(3)_f$ (f for 'flavour') of hadronic physics this way. This is, at least approximately, a manifest symmetry. We'd need 8 Yang–Mills vector quanta, which, in the absence of weak and electromagnetic corrections (taking the traditional view of these), would all be massless if $SU(3)_f$ were exact. Turning on the electromagnetic and weak interactions, the isospin 'charges' I_1 and I_2 fail to be conserved because of electromagnetic corrections, and so we would expect 'their' W's to have masses of order α times a typical hadronic mass. The hypercharge Y should only be broken weakly, and so its associated W should be even lighter. Charge Q is presumably exactly conserved, the associated quantum being the massless γ. The remaining four vector quanta should have typical hadronic masses, associated with the strong breaking of $SU(3)_f$. Thus, we expect a collection of eight vector bosons – four with hadronic masses, two with masses of order a few MeV (electromagnetic), one with a mass even smaller than that, and the photon. Such a pattern of vector boson masses bears almost no relation to the observed spectrum, and so this scheme for hadronic interactions fails.

Now, almost thirty years after the original paper by Yang & Mills, it is accepted that hadrons are composite objects, made of quarks and antiquarks. Just as the simple electromagnetic (gauge) interaction describes the forces between the charged constituents of atoms, while the interatomic force is complicated and, in principle, derivable from the inter-constituent force, so one may hope that the interquark force may be simple, the complicated interhadronic one being then a secondary manifestation of it. It would clearly be a beautiful generalisation of the electromagnetic case if the interquark force could be described by a gauge field theory. Perhaps, then, 'fundamental' particles would be those which interact by the exchange of gauge field quanta. In any case, the best (many would now say the only) candidate for a theory of the strong interactions between quarks is precisely of this type – quantum chromodynamics, or QCD.

QCD is a non-Abelian gauge theory, based on the local extension of the $SU(3)_C$ (colour) symmetry mentioned in Section 1.1. The quarks belong to the fundamental three-dimensional representation of $SU(3)_C$, and we write the quark fields as

$$\psi_i = \begin{pmatrix} q_i{}^{\mathrm{R}} \\ q_i{}^{\mathrm{B}} \\ q_i{}^{\mathrm{G}} \end{pmatrix} \tag{3.63}$$

with i running over the flavour index (u, d, s, c, . . .) and R, B, G standing for the colour labels. Global $SU(3)_C$ transformations are described by

$$\psi_i \to \psi_i' = \exp(-\mathrm{i}\theta \cdot \lambda/2)\psi_i, \tag{3.64}$$

where the θ_α ($\alpha = 1, 2, \ldots, 8$) parametrise the general $SU(3)$ matrix, and the eight matrices λ^α (each 3×3) are the Gell-Mann (1962) matrices, which generalise the $\boldsymbol{\tau}$ matrices of $SU(2)$. They obey the algebra (cf. (3.51))

$$\left[\frac{\lambda^\alpha}{2}, \frac{\lambda^\beta}{2}\right] = \mathrm{i} f_{\alpha\beta\gamma} \frac{\lambda^\gamma}{2}, \tag{3.65}$$

where the f's are the structure constants of $SU(3)$. Making the invariance (3.64) into a local one, $\theta \to \theta(x)$, will generate a dynamical theory whose Lagrangian is easily written down, after the foregoing preparation:

$$\mathcal{L} = \sum_{i=\text{flavour}} \bar{\psi}_i \left(\mathrm{i}\not{\partial} - m_i - g_s \frac{\lambda^\alpha}{2} \not{A}^\alpha\right) \psi_i - \tfrac{1}{4} F_{\mu\nu}{}^\alpha F^{\alpha\mu\nu}, \tag{3.66}$$

the covariant derivative (3.1) of QED having been generalised to

$$D^\mu = \partial^\mu + \mathrm{i} g_s \frac{\lambda^\alpha}{2} A^{\alpha\mu}. \tag{3.67}$$

There are now *eight* gauge fields $A^{\alpha\mu}$, transforming as the regular representation of $SU(3)$; their quanta are called *gluons*. The tensor $F_{\mu\nu}{}^\alpha$ is

$$F_{\mu\nu}{}^\alpha = \partial_\mu A_\nu{}^\alpha - \partial_\nu A_\mu{}^\alpha - g_s f_{\alpha\beta\gamma} A_\mu{}^\beta A_\nu{}^\gamma, \tag{3.68}$$

and g_s is the strong interaction gauge coupling constant. Often the symbol $\alpha_s = g_s{}^2/4\pi$ is introduced. We have included non-zero quark masses in (3.66), though very possibly they should be zero (see further Sections 6.8 and 7.4 for comments about fermion masses).

Actually, we have to make clear that (3.66) is *not* the complete Lagrangian of the quantum field theory QCD - it is essentially the 'classical field' version, suitable for lowest-order perturbation theory (tree graphs) only (the rules for tree graphs in QCD are given in Aitchison & Hey (1982), for example). When we discuss the quantisation of gauge theories in the following chapter, we shall see (Section 4.6) that 'gauge-fixing' (and, generally, 'ghost') terms have to be added to (3.66). A further subtlety is that an additional term of the form

$$\theta \epsilon^{\mu\nu\rho\sigma} F_{\mu\nu}{}^\alpha F_{\rho\sigma}{}^\alpha, \tag{3.69}$$

where θ is an arbitrary parameter, may also be included in $\mathcal{L}$. A discussion of the origin of this term, and of various possible responses to it, would take us too far afield - see, for example, Llewellyn Smith (1982*a*).

The development and application of QCD is now a vast subject, necessitating a separate book on its own. Recent reviews are contained in Ellis & Sachrajda (1980), and Llewellyn Smith (1980, 1982*a*).

4 Quantisation of vector fields: I Massless

It is abundantly clear, at this stage, that we must understand how to quantise massless vector fields, since they play such a fundamental role in the 'gauge principle' approach to dynamics. Unfortunately, there is no escaping the fact that this is a difficult and technical matter. Indeed, the derivation of the full Feynman rules for non-Abelian gauge theories was for many years an outstanding problem.

In all gauge theories, a major difficulty arises simply as a consequence of the fact that gauge fields are not uniquely specified by the equations of motion (and given initial conditions), but can always be subjected to a gauge transformation without affecting physical observables. In fact, the vector field $A^\mu(x)$ - to take the specific $U(1)$ case - does not really have four independent components. From classical electromagnetic theory, we know perfectly well that the free electromagnetic field has only two independent components (polarisations), not the four corresponding to all of the A^0 and $\mathbf{A}$ in A^μ. The number of components can be reduced by imposing suitable conditions on A^μ - but then we are involved with the problem of field quantisation subject to constraints. Even the classical Hamiltonian dynamics of such field systems is awkward; what is more, in the quantum case, as we shall see, such constraints cannot be imposed as operator conditions at all. There are also further difficulties. In formulating the constraint on A^μ, we have the choice of trying a Lorentz covariant condition, for example the *Lorentz condition*

$$\partial_\mu A^\mu = 0 \tag{4.1}$$

familiar in the classical field case, or a non-covariant condition, such as

$$\operatorname{div} \mathbf{A} = 0. \tag{4.2}$$

Conditions such as (4.1) or (4.2) are termed 'gauge-fixing conditions', and choosing one - or another condition - is called 'choosing a gauge'. Potentials such that (4.1) holds are said to be in the 'Lorentz gauge'; (4.2) is the Coulomb gauge. If we adopt the covariant gauge choice (4.1), then, as we shall see, the quanta associated with A^0 (in the quantised case) create states of negative norm, and we run the risk of violating unitarity. On the other hand, if we use

(4.2), we lose manifest covariance. These difficulties are, of course, well known in the electromagnetic case, and their resolution is well discussed in the books by Mandl (1959), Bjorken & Drell (1965), and Itzykson & Zuber (1980), for example. However, we shall include in the present chapter some of the reasonably straightforward aspects of the quantum theory of massless vector fields, in order to make this book self-contained. The quantisation of massive vector fields will be discussed in the following chapter.

4.1 The electromagnetic field

As we saw in Section 3.1, the classical Maxwell equations can be put in manifestly covariant form by introducing the four-vector potential $A^\mu(x)$ which satisfies the equations of motion

$$\Box A^\mu - \partial^\mu \partial_\nu A^\nu = j^\mu. \tag{4.3}$$

These equations may also be written as

$$\partial_\nu F^{\nu\mu} = j^\mu, \tag{4.4}$$

where

$$F^{\mu\nu} = \partial^\mu A^\nu - \partial^\nu A^\mu. \tag{4.5}$$

From (4.5), we can remind ourselves that this theory is manifestly invariant under the classical gauge transformation

$$A_\mu(x) \rightarrow A'_\mu(x) = A_\mu(x) + \partial_\mu \chi(x). \tag{4.6}$$

Thus, in the classical case, we can use this freedom to constrain A^μ so as to satisfy

$$\partial_\mu A^\mu = 0. \tag{4.7}$$

A^μ is now in the Lorentz gauge, and the field equations reduce to the simple form

$$\Box A^\mu = j^\mu, \tag{4.8}$$

in which the different components of A^μ are uncoupled from each other. We note that (4.7) and (4.8) are consistent if $\partial_\mu j^\mu = 0$ (current conservation).

As a first step towards the canonical quantisation of this theory, we ask what Lagrangian gives the free-field equations

$$\Box A^\mu - \partial^\mu \partial_\nu A^\nu = 0. \tag{4.9}$$

As we saw in Section 3.1, the answer is

$$\mathcal{L}_{\text{em}}{}^{\text{class}} = -\tfrac{1}{4} F_{\mu\nu} F^{\mu\nu}. \tag{4.10}$$

Consider now the quantisation of this theory. We attempt to calculate the 'momenta' canonically conjugate to each component of A^μ, and impose equal time commutation relations of the form (2.9)-(2.11). All goes well for the spatial components of A^μ, but for A^0 we find a difficulty: clearly,

$$\pi^0 = \frac{\partial \mathcal{L}_{\text{em}}{}^{\text{class}}}{\partial(\partial_0 A^0)} \tag{4.11}$$

vanishes, and we cannot impose the desired commutation relation.

There is a further, related, difficulty. In introductory books on quantum field theory, it is shown how the momentum space propagator for a free field is (proportional to) the inverse of the Fourier transform of the differential operator multiplying the field in the equation of motion - i.e. it is the free-field Green function. For example, the scalar field equation is

$$(\Box + m^2)\phi = 0, \tag{4.12}$$

which, in momentum space, is

$$(-k^2 + m^2)\phi = 0 \tag{4.13}$$

and the scalar propagator is $(k^2 - m^2)^{-1}$ (disregarding a factor of i). In the same way, we expect the A^μ-propagator to be the inverse of the operator in (4.9), namely of $-k^2 g^{\mu\nu} + k^\mu k^\nu$. By covariance, this must have the general form $Ag_{\nu\lambda} + Bk_\nu k_\lambda$, where

$$(-k^2 g^{\mu\nu} + k^\mu k^\nu)(Ag_{\nu\lambda} + Bk_\nu k_\lambda) = g_\lambda{}^\mu. \tag{4.14}$$

Multiplying out, we find

$$-k^2 A g_\lambda{}^\mu - Bk^2 k^\mu k_\lambda + Bk^2 k^\mu k_\lambda + Ak^\mu k_\lambda = g_\lambda{}^\mu, \tag{4.15}$$

which is impossible for any A. This wave operator therefore has *no* inverse.

Both these problems can be solved, it turns out, if we consider an alternative Lagrangian which yields the equation of motion for A^μ which would follow from (4.9) if (4.7) were imposed; namely the Lagrangian

$$\mathcal{L}_{\text{em}} = -\tfrac{1}{4}F_{\mu\nu}F^{\mu\nu} - \tfrac{1}{2}(\partial_\mu A^\mu)^2. \tag{4.16}$$

As can easily be checked, the Euler-Lagrange equation for A^μ, following from (4.16), is indeed

$$\Box A^\mu = 0. \tag{4.17}$$

It follows that $\Box(\partial \cdot A) = 0$, so that $\partial \cdot A$ is a free massless field. A current term or terms can of course always be added to (4.16) as well; $(\partial \cdot A)$ would still be free, if the current is conserved. With the modified Lagrangian (4.16), π^0 is now not zero, but is given by

$$\pi^0 = -\partial_\mu A^\mu \tag{4.18}$$

and, furthermore, the inverse of the operator in (4.17) is perfectly well defined, being simply

$$-g^{\mu\nu}/k^2. \tag{4.19}$$

However, there is a subtlety. We are not free to suppose that the Lorentz condition

$$\partial_\mu A^\mu = 0 \tag{4.20}$$

actually *has* been imposed as an operator condition, since it conflicts with the canonical commutation relations we have carefully obtained. These relations would read, for example,

$$[A_\mu(\mathbf{x}, t), \pi^\nu(\mathbf{x}', t)] = \mathrm{i} g_\mu{}^\nu \delta^3(\mathbf{x}-\mathbf{x}'). \tag{4.21}$$

Setting $\mu=\nu=0$, the right-hand side is certainly non-zero, yet the left-hand side vanishes if (4.18) and (4.20) both hold.

We shall return to (4.20) in a moment. First, we recall the conventional normal mode expansion of A^μ (treating now all components as independent degrees of freedom), based on the commutation relations (4.21), together with

$$[A_\mu(\mathbf{x}, t), A_\nu(\mathbf{x}', t)] = [\pi_\mu(\mathbf{x}, t), \pi_\nu(\mathbf{x}, t)] = 0. \tag{4.22}$$

(4.21) and (4.22) are, at first sight, just the expected generalisation of the scalar case (2.9)-(2.11). But there is actually a significant difference. From (4.22), it follows that the spatial derivatives of A_μ commute at equal times, and we may write (4.21) and (4.22) as

$$[A_\mu(\mathbf{x}, t), \dot{A}_\nu(\mathbf{x}', t)] = -\mathrm{i} g_{\mu\nu}\delta^3(\mathbf{x}-\mathbf{x}') \tag{4.23}$$

$$[\dot{A}_\mu(\mathbf{x}, t), \dot{A}_\nu(\mathbf{x}', t)] = 0 \tag{4.24}$$

$$[A_\mu(\mathbf{x}, t), A_\nu(\mathbf{x}', t)] = 0. \tag{4.25}$$

The spatial parts of A_μ, therefore (recalling $g_{11}=g_{22}=g_{33}=-1$) do obey the scalar field commutation relation (2.9), but the $\mu=0$ component has the 'wrong' sign. We shall see that this component is associated with states of negative norm. The normal mode expansion of A_μ is

$$A_\mu(x) = \int \frac{\mathrm{d}^3\mathbf{k}}{[(2\pi)^3 2\omega]^{1/2}} \sum_{\lambda=0}^{3} [a^{(\lambda)}\epsilon_\mu{}^{(\lambda)}(k)\,\mathrm{e}^{-\mathrm{i}k\cdot x} + a^{(\lambda)\dagger}\epsilon_\mu{}^{(\lambda)*}(k)\,\mathrm{e}^{\mathrm{i}k\cdot x}] \tag{4.26}$$

(cf. (2.60)), where $\omega = |\mathbf{k}|$ for the massless quanta. The particle interpretation follows from the commutation relations

$$[a^{(\lambda)}(k), a^{(\lambda')\dagger}(k')] = -g^{\lambda\lambda'}\delta^3(\mathbf{k}-\mathbf{k}'). \tag{4.27}$$

The 'wrong' sign appears in the case $\lambda=\lambda'=0$.

The polarisation vectors $\epsilon_\mu{}^{(\lambda)}(k)$ (for $\lambda=0, 1, 2, 3$) form a set of four mutually orthogonal unit vectors, whose choice is to some extent arbitrary; as the notation indicates, they may depend on k. A conventional possibility is to choose $\epsilon^{(1)}$, $\epsilon^{(2)}$ and $\epsilon^{(3)}$ to have vanishing time components, and $\epsilon^{(1)}$, $\epsilon^{(2)}$ to have space components perpendicular to $\hat{\mathbf{k}}=\mathbf{k}/|\mathbf{k}|$, while the space component of $\epsilon^{(3)}$ is along $\hat{\mathbf{k}}$; $\epsilon^{(0)}$ is then (1, 0, 0, 0). The vectors $\epsilon^{(1)}$ and $\epsilon^{(2)}$ describe states of transverse polarisation, $\epsilon^{(3)}$ a state of longitudinal polarisation, and $\epsilon^{(0)}$ a 'scalar' photon state.

The vacuum $|0\rangle$ must presumably be the state without any kind of photon:

$$a^{(\lambda)}(k)|0\rangle = 0; \quad \lambda = 0, 1, 2, 3. \tag{4.28}$$

However, we run into trouble if we try to compute the norm of a state with one scalar photon. Such a state is written as

$$|\gamma, \lambda = 0\rangle = \int \mathrm{d}^3\mathbf{k} f(k) a^{(0)\dagger}(k)|0\rangle, \tag{4.29}$$

where the wave packet function $f(k)$ satisfies the square-integrability condition

$$\int |f|^2 \, \mathrm{d}^3\mathbf{k} < \infty. \tag{4.30}$$

We find, using (4.27), that the state (4.29) has *negative* norm:

$$\langle\gamma, \lambda = 0|\gamma, \lambda = 0\rangle = -\langle 0|0\rangle \int |f(k)|^2 \, \mathrm{d}^3\mathbf{k}. \tag{4.31}$$

In the same way, the Hamiltonian turns out to be

$$H = \int \mathrm{d}^3\mathbf{k} \omega(-g_{\lambda\lambda'} a^{(\lambda)\dagger}(k) a^{(\lambda')}(k)), \tag{4.32}$$

in which the $\lambda' = \lambda = 0$ contribution is not positive definite.

The $\lambda = 0$ state is therefore definitely unwanted. So, however, is the $\lambda = 3$ state since, as we saw above, the free Maxwell field has really only two independent components, which are transverse to $\hat{\mathbf{k}}$. In the *classical* case, this fact is simply seen in terms of the Lorentz condition (4.20). This condition does not determine A_μ uniquely, since a gauge transformation (4.6) is still possible, provided

$$\Box \chi = 0. \tag{4.33}$$

For a plane wave describing a particle of zero rest mass, we may take

$$A_\mu(x) = \epsilon_\mu \, \mathrm{e}^{\pm i k \cdot x} \tag{4.34}$$

$$\chi(x) = c \, \mathrm{e}^{\pm i k \cdot x} \Bigg\} \; k^2 = 0, \tag{4.35}$$

and the gauge transformation (4.6) is

$$\epsilon_\mu \to \epsilon_\mu \pm i c k_\mu. \tag{4.36}$$

Thus ϵ_0 (for example) can always be reduced to zero, and condition (4.20) becomes

$$\mathbf{k} \cdot \boldsymbol{\epsilon} = 0, \tag{4.37}$$

expressing transversality.

This suggests that we might be able to use (4.20) in the quantum theory to get rid of the unwanted $\lambda = 0, 3$ states. However, we have seen that (4.20) cannot consistently be interpreted as an operator equation. In fact, though, we do not really need a condition as strong as (4.20): we only want something which forbids unwanted photons in physical states. This can be achieved (Mandl (1959), Itzykson & Zuber (1980)) by the weaker condition

$$\partial_\mu A_\mu{}^{(+)}(x)\,|0\rangle = 0 \tag{4.38}$$

imposed on the positive frequency (annihilating) part of $\partial \cdot A$, as applied to the physical vacuum. Using (4.26) and our choice of $\epsilon_\mu{}^{(\lambda)}$, (4.38) can be written as

$$[a^{(0)}(k) - a^{(3)}(k)]\,|0\rangle = 0. \tag{4.39}$$

According to (4.39), the vacuum is now not necessarily simply the state with no longitudinal or scalar photons. In fact, it is not uniquely defined by (4.39), but can have arbitrary admixtures of such photons, provided always that condition (4.39) is satisfied. Mandl (1959), and Itzykson & Zuber (1980), show that changing these admixtures corresponds exactly to a gauge transformation on A_μ, and has therefore no physical significance. Finally, then, the vacuum may, without loss of generality, be taken as in (4.28), for all k, and a consistent quantum theory set up, based on (4.28) and (4.39).

One may easily check, for example, that the condition (4.39) ensures that the energy of the longitudinal and scalar photons is zero (they cancel each other) in allowed states:

$$\begin{aligned}
&\langle 0|\,-a^{(0)\dagger}a^{(0)} + a^{(3)\dagger}a^{(3)}\,|0\rangle \\
&= \langle 0|(-a^{(0)\dagger} + a^{(3)\dagger})a^{(3)}|0\rangle \qquad \text{using (4.39)} \\
&= 0 \qquad \text{using the Hermitian conjugate of (4.39).}
\end{aligned}$$

The mean value of the energy (and momentum) therefore only involves transverse photons.

The photon propagator (4.19) originating from (4.17) is said to be in 'Feynman gauge'.

4.2 Other gauges

We have seen that the Lagrangian (4.16) gives the field equations for A^μ in a form which would have resulted from imposing $\partial \cdot A = 0$ on (4.9). Such a modification to the Maxwell Lagrangian (4.10) seemed to be necessary, in view of the difficulties with π^0 and with the photon propagator. However, this modification is by no means the only one which circumvents these difficulties. Indeed, renormalisation (higher-order) effects will, in general, modify the numerical coefficients of the two terms in (4.16) anyway. This leads us to consider the more general Lagrangian

$$\mathcal{L}_{\text{em}} = -\tfrac{1}{4}F_{\mu\nu}F^{\mu\nu} - \frac{1}{2\xi}(\partial \cdot A)^2. \tag{4.40}$$

The equation of motion is now

$$\left(\Box g^{\mu\nu} + \left(\frac{1}{\xi} - 1\right) \partial^\mu \partial^\nu\right) A_\nu = 0 \tag{4.41}$$

and $\partial \cdot A$ is still a free field. π^0 is given by

$$\pi^0 = -\frac{1}{\xi}(\partial \cdot A). \tag{4.42}$$

The momentum space operator in (4.41) is

$$\left(-k^2 g^{\mu\nu} + \left(1 - \frac{1}{\xi}\right) k^\mu k^\nu\right), \tag{4.43}$$

which has the inverse (i.e. the photon propagator)

$$(-g^{\mu\nu} + (1 - \xi)k^\mu k^\nu/k^2)/k^2. \tag{4.44}$$

The ξ term in (4.4) is called the 'gauge-fixing' term. We note that the Feynman gauge choice is recovered as $\xi \to 1$, but that the limit $\xi \to \infty$ (no gauge-fixing term) does not exist. Another interesting choice is the Landau gauge, $\xi = 0$. From (4.44), this produces a photon propagator whose scalar product with k^μ is always zero (i.e. the propagator is transverse to k^μ).

The case $\xi = 0$ is, however, singular in the Lagrangian (corresponding to the fact that the propagator (4.44) for $\xi = 0$ is non-invertible). An alternative formulation is possible (Nakanishi 1966, 1973, 1974; Lautrup 1967) which includes this special case more satisfactorily. Consider the Lagrangian

$$\mathcal{L}_{\text{em}} = -\tfrac{1}{4}F_{\mu\nu}F^{\mu\nu} + B\partial \cdot A + \tfrac{1}{2}\xi B^2, \tag{4.45}$$

where $B(x)$ is a scalar field. The momentum conjugate to A^0 is now

$$\pi^0 = B, \tag{4.46}$$

while the Euler-Lagrange equations for A_μ yield

$$\Box A^\mu - \partial^\mu \partial_\nu A^\nu = \partial^\mu B \tag{4.47}$$

and for B give

$$\partial \cdot A + \xi B = 0. \tag{4.48}$$

Eliminating B from (4.47) by means of (4.48), we recover (4.41). From (4.47) we learn that $\Box B = 0$, and from (4.48) that $\Box(\partial \cdot A) = 0$, so that both B and $\partial \cdot A$ are free. We note that (4.45) can also be written as

$$\mathcal{L}_{\text{em}} = -\tfrac{1}{4}F_{\mu\nu}F^{\mu\nu} + \frac{1}{2\xi}(\partial \cdot A + \xi B)^2 - \frac{1}{2\xi}(\partial \cdot A)^2, \tag{4.49}$$

in which the combination $C=(\partial\cdot A+\xi B)$ has an Euler–Lagrange equation simply $C=0$; it is a decoupled free field.

In this formalism, the appropriate subsidiary condition for getting rid of unphysical states is

$$B^{(+)}(x)|0\rangle=0. \tag{4.50}$$

Kugo & Ojima (1979) show that (4.50) provides a satisfactory definition of the Hilbert space of physical states.

Of course, it is essential to prove that all physical results are independent of the choice of the gauge parameter ξ. We shall discuss this further in Section 4.4.

In closing this section we should note once again that we have considered only *covariant* gauges. Other, non-covariant, gauges are also possible, and often convenient, as will be mentioned at the end of this chapter.

4.3 The problem with non-Abelian gauge theories

We may attempt to generalise the foregoing in a straightforward way, starting from the Lagrangian introduced in (3.47)

$$\mathcal{L}_W=-\tfrac{1}{4}\mathbf{F}_{\mu\nu}\cdot\mathbf{F}^{\mu\nu}, \tag{4.51}$$

the non-Abelian analogue of (4.10), for the special case of an $SU(2)$ group. In (4.51), we recall that $\mathbf{F}^{\mu\nu}$ is

$$\mathbf{F}^{\mu\nu}=\partial^\mu\mathbf{W}^\nu-\partial^\nu\mathbf{W}^\mu-g\mathbf{W}^\mu\times\mathbf{W}^\nu. \tag{4.52}$$

Once again, we have a gauge-invariance, and the freedom to transform $\mathbf{W}^\mu$ by a gauge transformation of the form (3.36). We need to consider a gauge-fixing procedure. The obvious analogue of (4.45) would be the addition of a term

$$\mathbf{B}\cdot(\partial\cdot\mathbf{W})+\tfrac{1}{2}\xi\mathbf{B}\cdot\mathbf{B} \tag{4.53}$$

to $\mathcal{L}_W$, where $\mathbf{B}$ transforms as the regular representation of $SU(2)$, like $\mathbf{W}^\mu$. (4.53) gives

$$\partial\cdot\mathbf{W}+\xi\mathbf{B}=\mathbf{0} \tag{4.54}$$

as the gauge condition, and

$$D^{ab\nu}F_{\mu\nu}{}^b+\partial_\mu B^a=0 \tag{4.55}$$

as the equation of motion for $\mathbf{W}_\mu$ ($a,b=1,2,3$; the covariant derivative in (4.55) is of course the one appropriate to a field in the regular representation – see comment 4 at the end of Section 3.3). It is possible to verify that

$$D^{ca\mu}D^{ab\nu}F_{\mu\nu}{}^b=0, \tag{4.56}$$

which implies that

$$D^{ca\mu}\partial_\mu B^a=0, \tag{4.57}$$

showing that the auxiliary field **B** is *not* a free field, in this non-Abelian case. Neither, therefore, is $\partial \cdot \mathbf{W}$. In consequence, the obvious generalisations of (4.38) and (4.50) cannot be used to define the physical states. Finding the correct generalisations of these conditions was, for many years, a major problem in the canonical quantisation of non-Abelian gauge theories.

In order to motivate the solution to this problem, we shall return to the Abelian (electromagnetic) case, this time focussing attention on the role of *Ward identities* in unitarity.

4.4 Unitarity and Ward identities: loop graphs in QED

Despite the learned-sounding heading, we are still not going to be very formal. The problem we are going to discuss is again the elimination of the unphysical polarisation states, but this time we shall approach it from a different angle – that of *unitarity*. We said in Section 4.2, of course, that the $\lambda = 0$ states had negative norm, so that we have cause to worry about conservation of probability. A fuller discussion of some parts of the argument is given in Aitchison & Hey (1982); some overlap with the present treatment is unavoidable.

Consider any scattering amplitude whose imaginary part has a contribution from an intermediate state involving just one photon (we do not need to specify what else the state contains, at this stage), as shown in Fig. 4.1. We write, for just this contribution,

$$\mathrm{Im}\,\langle f|F|i\rangle = \int \sum_n \langle f|G|n\rangle\langle n|G^\dagger|i\rangle d\rho_n, \tag{4.58}$$

where $d\rho_n$ is the phase space element for the state. A one-photon amplitude will have the form

$$\epsilon_\mu{}^{(\lambda)}(k)G^\mu, \tag{4.59}$$

where now, since the photon is a *physical* particle in the unitarity relation (4.58), only two values of λ are allowed, say $\lambda = 1, 2$, corresponding to transversely polarised photons. The sum in (4.58) is then over these two polarisation states, so that (suppressing the i, f labels)

$$\mathrm{Im}\,F = \int G^\mu \left[\sum_{\lambda=1,2} \epsilon_\mu{}^{(\lambda)}(k)\epsilon_\nu{}^{(\lambda)}(k) \right] G^{\nu\dagger} d\rho_n, \tag{4.60}$$

Fig. 4.1. Contribution to the imaginary part of a $2 \rightarrow 2$ amplitude, involving one photon in the intermediate state.

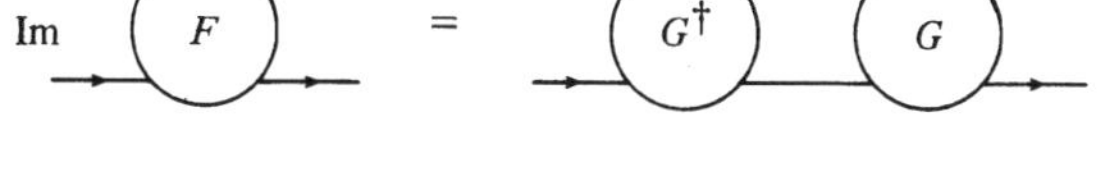

where we have adopted real polarisation vectors, and $k^2 = 0$ for the physical photons. With $k = (|\mathbf{k}|, 0, 0, |\mathbf{k}|)$, we take $\epsilon^{(1)} = (0, 1, 0, 0)$, $\epsilon^{(2)} = (0, 0, 1, 0)$.

Now consider a Feynman graph in which a single photon appears inside a loop - for example, the box graph shown in Fig. 4.2. Such a graph may be thought of as a specific contribution to the process $i \to f$ - one in which, in particular, there is just one photon in the intermediate state, together with another particle called A. The question we must ask is this: does the imaginary part of the amplitude for Fig. 4.2, as calculated according to the rules for propagators and vertices as so far developed, really have the same form as (4.58)?

To be specific, let us consider Feynman gauge. The photon propagator is then very simple, as we have seen:

$$-g_{\mu\nu}/(k^2 - \mathrm{i}\epsilon), \tag{4.61}$$

where we have now included the infinitesimal imaginary part in the denominator, as we are interested precisely in the imaginary part of the box graph. As with all loops, this graph will involve a four-dimensional integral over k, and the imaginary part will come from the contribution of the pole in (4.61) at $k^2 = 0$. The rule for obtaining the imaginary part is (Cutkosky 1960) to replace (4.61) by

$$-g_{\mu\nu}\pi\delta(k^2)\theta(k_0), \tag{4.62}$$

i.e. the photon is put on-mass-shell with positive energy. Similarly, the other intermediate state particle, A, is put on-shell. These two conditions have the effect of converting the integral over k (in the full box graph) into the standard two-body phase space integral for the $\gamma + A$ state, so that if we call the box amplitude B, we have finally

$$\operatorname{Im} B = \int T^{\mu}(-g_{\mu\nu})T^{\nu\dagger} d\rho_2, \tag{4.63}$$

where T^{μ} is the tree amplitude (with all particles on-shell) involving one photon leg, as shown in Fig. 4.3.

Now (4.63) certainly has the same general appearance as (4.60), and looks like a 'verification' of (4.60) to lowest order in the coupling strength e (the box graph on the left-hand side is of order e^2, and the two tree graphs are each of

Fig. 4.2. Box graph contribution to F.

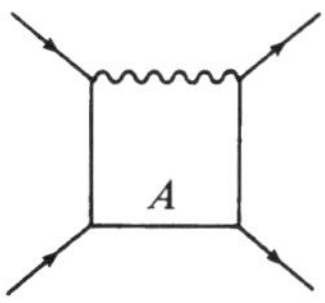

order e). However, there is a crucial difference. The photon in (4.63), despite having $k^2 = 0$ and positive energy, according to (4.62), is still not completely physical, as it is in (4.60). It is simple to verify, using the ϵ's introduced in Section 4.1, that the factor $-g_{\mu\nu}$ in (4.63) may be regarded as a photon polarisation sum, but one in which all *four* polarisation states enter

$$-\epsilon_\mu^{(0)}(k)\epsilon_\nu^{(0)}(k) + \sum_{\lambda=1}^{3} \epsilon_\mu^{(\lambda)}(k)\epsilon_\nu^{(\lambda)}(k) = -g_{\mu\nu}. \tag{4.64}$$

In (4.60), of course, only the transverse states $\lambda = 1, 2$ appear - that is to say, (4.60) includes only those photon states for which $\epsilon \cdot k = 0$ *as well as* $k^2 = 0$. If $k = (|\mathbf{k}|, 0, 0, |\mathbf{k}|)$, $\epsilon_\mu^{(0)}(k)$ and $\epsilon_\mu^{(3)}(k)$ do *not* satisfy $\epsilon \cdot k = 0$. Thus unwanted unphysical contributions are apparently present in (4.63).

To see if these contributions matter, we need an expression for the difference between

$$P_{\mu\nu} = \sum_{\lambda=1,2} \epsilon_\mu^{(\lambda)}(k)\epsilon_\nu^{(\lambda)}(k) \tag{4.65}$$

and the corresponding quantity, which enters in (4.64). As discussed, for example, in Taylor (1978) and in Aitchison & Hey (1982), two transverse polarisation vectors are not uniquely defined by the orthogonality condition

$$\epsilon_\mu^{(\lambda)}(k)\epsilon^{(\lambda')\mu}(k) = -\delta_{\lambda\lambda'} \tag{4.66}$$

and the Lorentz condition $k \cdot \epsilon = 0$, all with $k^2 = 0$. This is because a gauge transformation on the potentials can always be done, leading to a change in the ϵ's of the form (4.36), and these new ϵ's satisfy the stated conditions just as well as the old ones. To make the ϵ's unique, one can introduce a further condition

$$t \cdot \epsilon = 0, \tag{4.67}$$

where t_μ is some constant four-vector. One then finds that $P_{\mu\nu}$ is given by

$$P_{\mu\nu} = -g_{\mu\nu} - [t^2 k_\mu k_\nu - k \cdot t(k_\mu t_\nu + k_\nu t_\mu)]/(k \cdot t)^2. \tag{4.68}$$

It follows, finally, that the unwanted contributions in (4.63), involving $-g_{\mu\nu} - P_{\mu\nu}$, will vanish provided

$$k_\mu T^\mu = 0. \tag{4.69}$$

Fig. 4.3. Tree graph entering into the imaginary part of B.

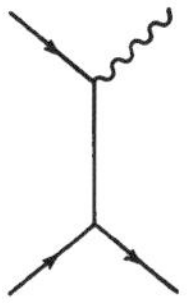

Equation (4.69) is a *Ward identity*, when T^μ is generalised to a complete amplitude G^μ involving one external photon: it states that if, in the amplitude $\epsilon_\mu G^\mu$, ϵ_μ is replaced by k_μ, the result is zero. We note that, as far as $P_{\mu\nu}$ itself is concerned, (4.68) ensures that the dependence on the arbitrary four-vector t_μ will disappear - which is just as well, since otherwise Lorentz covariance would have been threatened. Furthermore, (4.69) also allows us to use the more general ξ-dependent propagator (4.44), since the extra bits are merely proportional to k.

For the 'one-photon' amplitude we have been considering, (4.69) is actually just a statement of current conservation, as the following argument (Taylor 1978) shows. An amplitude of the type G^μ, involving one external photon, will be proportional to the Fourier transform

$$\int d^4x\, e^{ik\cdot x}\langle A \text{ out } | j_{em}{}^\mu(x) | B \text{ in}\rangle, \tag{4.70}$$

where A and B stand for all the other legs, apart from the photon. By a partial integration, $k_\mu G^\mu$ can be related to the divergence $\langle A \text{ out } | \partial_\mu j_{em}{}^\mu(x) | B \text{ in}\rangle$, which vanishes by current conservation.

It is tempting to conclude that unphysical polarisation states will always vanish from imaginary parts of loop graphs, simply because of current conservation. This is slightly too hasty, however. Consider, following Taylor (1978), a unitarity summation involving *two* photons, as in Fig. 4.4. Let the four-momenta and polarisations of the two photons be ϵ_1, k_1 and ϵ_2, k_2. We must now eliminate from Im F the unphysical components of the polarisation vectors of *both* photons, and thus require

$$k_{1\mu}T^{\mu\nu} = k_{2\nu}T^{\mu\nu} = 0, \tag{4.71}$$

where $T^{\mu\nu}$ is the two-photon amplitude of Fig. 4.5. $T^{\mu\nu}$ is related to the Fourier

Fig. 4.4. Contribution to an imaginary part which involves two photons in the intermediate state.

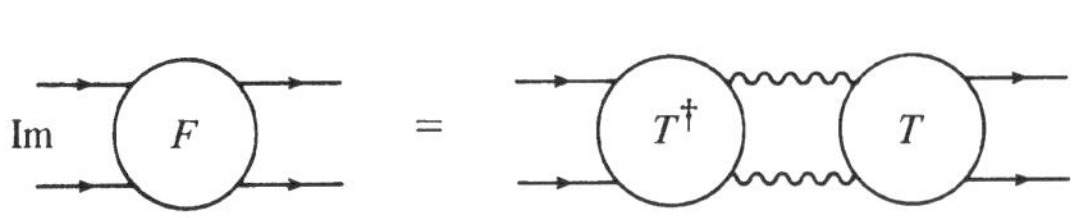

Fig. 4.5. Two-photon amplitude.

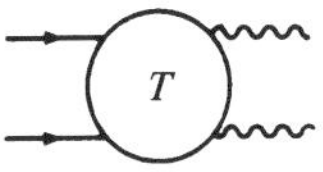

transform of

$$\langle A \text{ out } | T(j_{\text{em}}{}^{\mu}(x) j_{\text{em}}{}^{\nu}(y)) | B \text{ in} \rangle, \tag{4.72}$$

where T is the time-ordering operator. Consider the first of conditions (4.71). In this case, as Taylor (1978) points out, we have†

$$\partial_\mu T(j_{\text{em}}{}^{\mu}(x) j_{\text{em}}{}^{\nu}(y)) = T(\partial_\mu j_{\text{em}}{}^{\mu}(x) j_{\text{em}}{}^{\nu}(y)) + \delta(x_0 - y_0)[j_{\text{em}}{}^{0}(x), j_{\text{em}}{}^{\nu}(y)]. \tag{4.73}$$

The first term on the right-hand side of (4.73) does indeed vanish by current conservation, but the second does not. However, this equal time commutator is zero if the currents are constructed from fermion fields (as in (2.89)), while in the case of boson fields 'seagull' terms have to be added to $T^{\mu\nu}$ which cancel Schwinger terms in the commutator (Jackiw 1972; see also Section 8.2 below). Thus the upshot is that (4.71) *is* true in QED, and guarantees the disappearance of the unphysical polarisation states.

It is interesting to consider (4.71) in a tree approximation, as we did for the one-photon process. Let us take the amplitude $e^+e^- \to \gamma\gamma$ as an illustration. Here, in lowest order, there are *two* tree graphs, shown in Figs 4.6 and 4.7. Using the standard QED rules one can easily check that (4.71) does hold for the sum of these two graphs (though not for either one separately).

4.5 More troubles with the non-Abelian case

We now enquire how much of the foregoing can be carried over to a non-Abelian gauge theory. The propagators for the gauge field quanta have just the

Fig. 4.6. Tree graph contribution to $e^+e^- \to \gamma\gamma$ (direct).

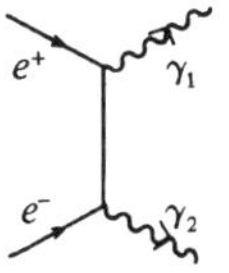

Fig. 4.7. Tree graph contribution to $e^+e^- \to \gamma\gamma$ (crossed).

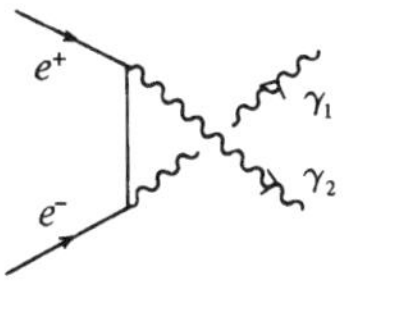

† Results similar to this will be derived in Section 8.2, where a fuller discussion of Ward identities will be given.

same form as (4.44), except for a factor of the form δ^{ab}, expressing the fact that they are diagonal in the internal quantum numbers. Consequently, exactly the same problem arises with unphysical polarisation states in unitarity summations. In this non-Abelian case, however, these are new complications. This may be seen in two ways. In the first place the currents, as we have seen in Section 3.3 (comment 8), now carry an internal space index, and even if they are conserved, commutators of the form

$$\delta(x_0 - y_0)[j_0{}^a(x), j_\nu{}^{b\dagger}(y)], \tag{4.74}$$

such as would appear in the non-Abelian analogue of (4.73), will not vanish. Secondly, one can see whether conditions (4.71) hold in the non-Abelian case at the tree graph level - for example, for the Born amplitudes for $q\bar{q} \to gg$ shown in Fig. 4.8. The details are given in Aitchison & Hey (1982), but the result is that (4.71) does *not* hold. When, in the sum of these three graphs, the polarisation vector of *one* gluon is replaced by its four-momentum, a part of (the characteristically non-Abelian) Fig. 4.8 (c) remains, which only vanishes if the *other* gluon is transverse. But in one-loop graphs constructed from Figs. 4.8, non-transverse gluon states will appear, just as in (4.64). Thus loop graphs involving two (or more) non-Abelian quanta will violate unitarity, if they are constructed according to the simple rules for tree graphs which follow from the Lagrangians of Sections 3.3 and 3.4.

Thus the Feynman rules for *loops* in non-Abelian gauge field theory are not straightforward. A heuristic approach would be to postulate the existence of an extra term (or terms) in the Lagrangian, which would have the sole function of cancelling off the unwanted polarisation states. Some idea of the form of such terms can be gained by considering what is needed in the specific case of the gluon loop $q\bar{q} \to gg \to q\bar{q}$, shown in Fig. 4.9. This sort of reasoning was first

Fig. 4.8. Tree graphs in $q\bar{q} \to gg$.

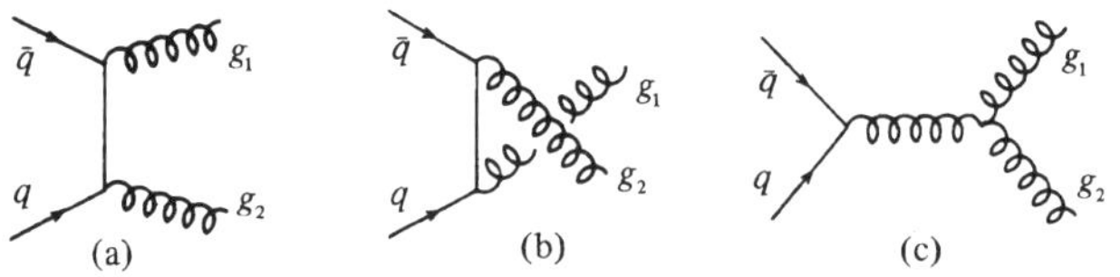

Fig. 4.9. Gluon loop contribution to $q\bar{q} \to q\bar{q}$.

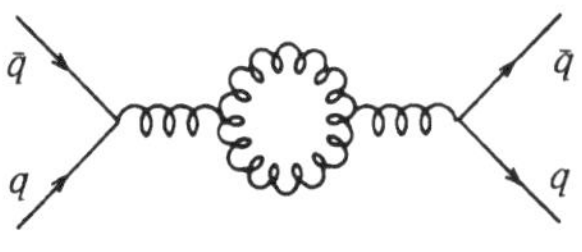

considered by Feynman (1963), and is given in some detail in Aitchison & Hey (1982), but it is difficult to generalise to arbitrary loops. A compact derivation of the correct rules for forming unitarity-preserving amplitudes in non-Abelian gauge theories was given by Faddeev & Popov (1967), using a path-integral formalism for quantisation.† They showed that extra terms had indeed to be added to the (gauge-fixed) Lagrangian of (4.51) together with (4.53). These terms had the very curious property of involving extra fields which were apparently Lorentz scalar, yet which obeyed fermionic commutation relations (essentially, in order to produce the *minus* sign required to cancel the remaining unwanted part in processes such as Fig. 4.9). Such fields are called 'ghosts'. Although their contributions do restore unitarity - as was proved by 't Hooft (1971*a*, *b*) - it should be emphasised that they do not represent physical particles; the precise form of the ghost terms depends on the choice of gauge, as we shall see in the following section - indeed it is possible to choose gauges with no ghosts at all.

Although it is certainly true that the path-integral formalism now offers the quickest route to the Feynman rules for gauge theories, it is more in keeping with the informal nature of the present introduction to the subject to stay with the canonical quantisation approach on which we embarked in Section 4.3, and to which we now return.

4.6 Canonical quantisation of non-Abelian gauge theories

We shall certainly not attempt a complete treatment, but will try to show how the ghost terms are introduced. Once again, we start with QED. In the preceding section, we learned that it is the Ward identities (4.69) or (4.71) (and their generalisations to processes involving more photons) that guarantee unitarity in QED. They are also crucial in the proof of renormalisability of QED (see Chapter 8 and Taylor 1978). We may therefore reasonably enquire what feature of the QED Lagrangian guarantees these identities. Now it is undoubtedly true that *if* our theory is invariant under the transformation

$$\epsilon_\mu \to \epsilon_\mu + ck_\mu, \tag{4.75}$$

then an amplitude of the form $\epsilon_\mu G^\mu$ is unchanged only if

$$k_\mu G^\mu = 0. \tag{4.76}$$

However, is full gauge-invariance of the Lagrangian *necessary* for such Ward identities? The answer had better be no, since the gauge-fixed Lagrangians we have advocated for the radiation field, (4.16) or (4.40) or (4.45), are certainly

† The Faddeev–Popov procedure is presented by Abers & Lee (1973), Taylor (1978), and Itzykson & Zuber (1980).

not invariant under

$$A_\mu \to A_\mu + \partial_\mu \chi! \qquad (4.77)$$

The true situation is that, although these QED Lagrangians (with matter contributions included) are not invariant under (4.77), there is some invariance left, which is actually enough to derive the required Ward identities.

Returning to Section 4.3, we note that the condition (4.20), though it constrains A_μ, still does not fix it uniquely. A_μ can change by $\partial_\mu \omega(x)$, say, and still satisfy $\partial \cdot A = -\xi B$, provided $\Box \omega = 0$. This is analogous to the point we made in Section 4.1, above equation (4.33). Thus the Lagrangian of (4.45),

$$\mathcal{L}_{\text{em}} = -\tfrac{1}{4} F_{\mu\nu} F^{\mu\nu} + B \partial \cdot A + \tfrac{1}{2} \xi B^2 \qquad (4.78)$$

is invariant under this restricted class of gauge transformations. The restriction on ω may be introduced into $\mathcal{L}_{\text{em}}$ via a multiplier field $\eta(x)$, leading to

$$\mathcal{L}_{\text{em}} = -\tfrac{1}{4} F_{\mu\nu} F^{\mu\nu} + B \partial \cdot A + \tfrac{1}{2} \xi B^2 - \eta \Box \omega, \qquad (4.79)$$

since then the Euler-Lagrange equation (2.5) for the η field yields just $\Box \omega = 0$.

The Lagrangian (4.79) is invariant under the (infinitesimal) transformation

$$A_\mu(x) \to A_\mu(x) + \epsilon \partial_\mu \omega(x) \qquad (4.80)$$

$$\eta(x) \to \eta(x) + \epsilon B(x) \qquad (4.81)$$

$$\omega(x) \to \omega(x) \qquad (4.82)$$

$$B(x) \to B(x) \qquad (4.83)$$

(were matter fields to be included, an appropriate transformation for them could be defined). The role of the η-ω term in (4.79) is to cancel the change in the gauge-fixing term, when A_μ undergoes the transformation (4.80). Thanks to this cancellation, a *type* of 'gauge-invariance' is restored - namely, that displayed in (4.80)-(4.83). It is a peculiar kind of invariance, however: the transformations are parametrised by the x-independent global parameter ϵ, yet (4.80) also has a local aspect via the presence of $\omega(x)$.

The transformations (4.80)-(4.83) are called the (Abelian) BRS transformations (Becchi, Rouet & Stora 1974). The symmetry of $\mathcal{L}_{\text{em}}$ under these transformations is called BRS symmetry. It turns out (Brandt 1976; Levy 1979; Taylor 1978) that this (infinitesimal) BRS symmetry is all that is required to guarantee the Ward identities in QED; hence, (4.79) is a satisfactory Lagrangian. Actually, since the η and ω fields decouple from everything else, they can be ignored in practical calculations, leaving (4.45) or (4.40) as acceptable Lagrangians.

By way of introduction to the non-Abelian case, we may note that the η and ω fields do not necessarily always decouple; whether they do or not depends on the type of gauge chosen. In particular, if a 'non-linear' choice such as ('t Hooft

& Veltman 1972*b*)

$$\partial \cdot A = \lambda A^2 \tag{4.84}$$

were to be adopted, and introduced into $\mathcal{L}_{\text{em}}$ via a term

$$B(\partial \cdot A - \lambda A^2) + \tfrac{1}{2}\xi B^2, \tag{4.85}$$

we should find a first-order variation in the gauge-fixing term of amount equal to

$$\epsilon B(\Box\, \omega - 2\lambda A \cdot \partial\omega). \tag{4.86}$$

This has to be cancelled by the η–ω terms, and here, because of the $A \cdot \partial\omega$ coupling in (4.86), they will *not* be decoupled. What this means is that, were we to use the gauge (4.84), Feynman graphs for loops in QED would violate unitarity unless additional 'ghost' contributions were added.

This of course is just what we said had to happen in the non-Abelian case, which we now consider from the above viewpoint (again, thinking explicitly of the $SU(2)$ case). Certainly the gauge-fixed Lagrangian

$$\mathcal{L}_W = -\tfrac{1}{4}\mathbf{F}_{\mu\nu} \cdot \mathbf{F}^{\mu\nu} + \mathbf{B} \cdot \partial\mathbf{W} + \frac{\xi}{2}\mathbf{B} \cdot \mathbf{B} \tag{4.87}$$

is not gauge-invariant, when $\mathbf{W}_\mu$ changes by (3.36). We try to salvage a reduced gauge-invariance by introducing auxiliary fields $\eta(x)$ and $\omega(x)$ in such a way as to cancel off the variation of the gauge-fixing term under

$$\mathbf{W}_\mu(x) \to \mathbf{W}_\mu(x) + \partial_\mu\omega(x) + g\omega(x) \times \mathbf{W}_\mu(x). \tag{4.88}$$

The η–ω term can be guessed by considering the condition ω must satisfy for (4.54) still to hold after (4.88): we require

$$\partial^\mu(\partial_\mu\omega(x) + g\omega(x) \times \mathbf{W}_\mu(x)) = 0 \tag{4.89}$$

or

$$\partial^\mu D_\mu{}^{ab}\omega^b(x) = 0. \tag{4.90}$$

In this case, ω is not a free field (compare the previous $U(1)$ - QED - case, and also the comments after (4.86)), due to the self-coupling aspect of the Yang-Mills field. Condition (4.90) can be introduced via a multiplier field $\eta(x)$ leading to

$$\mathcal{L}_W = -\tfrac{1}{4}\mathbf{F}_{\mu\nu} \cdot \mathbf{F}^{\mu\nu} + \mathbf{B} \cdot \partial\mathbf{W} + \frac{\xi}{2}\mathbf{B} \cdot \mathbf{B} - \eta^a\partial^\mu D_\mu{}^{ab}\omega^b, \tag{4.91}$$

in which the ghost part can be alternatively written as $(\partial^\mu\eta^a)D_\mu{}^{ab}\omega^b$, after a partial integration.

The Lagrangian (4.91) does possess a residual invariance - the non-Abelian BRS symmetry - but it is naturally more complicated than in the $U(1)$ case. Let us suppose that while $\mathbf{W}_\mu$ transforms by

$$W_\mu{}^a \to W_\mu{}^a + \epsilon D_\mu{}^{ab}\omega^b, \tag{4.92}$$

B transforms by

$$B^a \to B^a \tag{4.93}$$

and η by

$$\eta^a \to \eta^a + \epsilon B^a. \tag{4.94}$$

The change due to (4.94) in the last term in (4.91) will then indeed cancel the variation in $\mathbf{B} \cdot \partial \mathbf{W}$, if $\omega^a \to \omega^a$ as in (4.82). But this last term also involves $\mathbf{W}_\mu$ (via the covariant derivative), and the change in $\mathbf{W}_\mu$ from (4.92) will result in a non-vanishing change in $\mathcal{L}_W$. To cancel it, a non-trivial transformation of ω^a is required, namely,

$$\omega^a \to \omega^a + \frac{\epsilon g}{2} \epsilon^{abc} \omega^b \omega^c. \tag{4.95}$$

For this change in ω not to vanish, ω must anticommute with itself, and for consistency in (4.95) we also need the C-number parameter ϵ to anticommute with ω. With these provisos (which sufficiently indicate the peculiar character of the auxiliary fields), (4.91) is invariant under the BRS transformations (4.92)-(4.95); and this is sufficient to prove all the required Ward identities.

A final requirement is the non-Abelian generalisation of the physical-space condition (4.50) - recall the discussion after equation (4.57). Kugo & Ojima (1979) have shown that it can be found by making use of the generator Q_B of the (one-parameter) BRS symmetry. Q_B can be identified via the Noether current $j_\mu{}^B$ corresponding to the transformations (4.92)-(4.95). Remarkably enough, the physical vacuum can be defined by

$$Q_B |0\rangle = 0. \tag{4.96}$$

Kugo & Ojima (1979) show that (4.96) reproduces (4.50) in the $U(1)$ case.

We conclude this discussion by giving the form of the gauge boson - ghost vertex which follows from (4.91), and which supplements the simple vertices obtainable from the Lagrangians of Chapter 3. Referring to Fig. 4.10, the ghost vertex is (for $SU(2)$)

$$-g\epsilon_{abc}k_\mu, \tag{4.97}$$

Fig. 4.10. Gauge boson - ghost vertex.

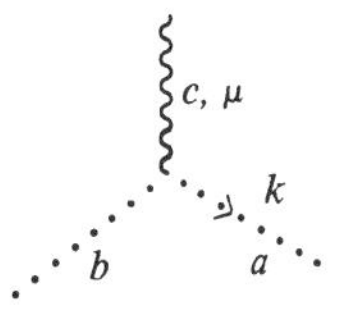

where k is the momentum of the outgoing ghost line. It is important to realise that the form of the ghost coupling depends, in general, on the type of gauge condition chosen. 't Hooft & Veltman (1972*b*) give a general prescription for obtaining the ghost Lagrangian corresponding to an arbitrary choice of gauge.

For completeness, we mention finally two other possible gauge choices. The first is the (non-covariant) Coulomb gauge defined (for the $U(1)$ case) by

$$\partial \cdot A - (t \cdot \partial)(t \cdot A) = 0, \qquad t^2 = 1. \tag{4.98}$$

(4.98) reduces to $\operatorname{div} \mathbf{A} = 0$ for $t = (1, 0, 0, 0)$. Non-Abelian indices can be added if desired. If we incorporated (4.98) by the addition of a gauge-fixing term $-\frac{1}{2}(\nabla \cdot \mathbf{A})^2$ in $\mathcal{L}_{\text{em}}{}^{\text{class}}$, we should still find $\pi^0 = 0$, and, by the same token, the wave operator for the A^μ fields is not invertible. However, quantisation in this gauge can be carried through (see, for example, Bjorken & Drell 1965), by accepting $\pi^0 = 0$ - and hence the loss of covariance - and by setting $\operatorname{div} \mathbf{A} = 0$ from the start, and thus quantising only the physical components of the field. In this case, the vector propagator is

$$\{-g^{\mu\nu} - [k^2 t^\mu t^\nu + k^\mu k^\nu - (k^\mu t^\nu + k^\nu t^\mu)] / [(k \cdot t)^2 - t^2 k^2]\}/k^2. \tag{4.99}$$

Bjorken & Drell (1965) show how the effect of the terms other than the $-g^{\mu\nu}$ part is to cancel the instantaneous Coulomb interaction (provided always that the current is conserved, so that the k^μ-dependent parts vanish), leaving effectively the Feynman gauge propagator, and thus restoring Lorentz covariance.

Another interesting possibility is the addition of a gauge-fixing term (Dokshitzer et al. 1980)

$$-\frac{1}{2\xi}(t \cdot A) \square (t \cdot A)\frac{1}{t^2}, \tag{4.100}$$

where $t^2 < 0$. These gauges are also not manifestly Lorentz covariant, since they depend on the four-vector t^μ. (4.100) yields the vector propagator

$$X^{\mu\nu} = \{-g^{\mu\nu} - [t^2(1-\xi)k^\mu k^\nu - (k \cdot t)(k^\mu t^\nu + k^\nu t^\mu)]/(k \cdot t)^2\}/k^2. \tag{4.101}$$

For $\xi \neq 0$, π^0 exists and (4.101) is invertible. The case $\xi = 1$ is called the 'planar' gauge. For $\xi \to 0$, the propagator (4.101) is in the so-called 'axial gauge', and satisfies

$$t_\mu X^{\mu\nu} = 0. \tag{4.102}$$

We note that, as expected from (4.102), the numerator of (4.101) for $\xi \to 0$ is just $P^{\mu\nu}$ of (4.68); it is then not invertible. The interpretation of the singularities at $(k \cdot t) = 0$ is discussed by Dokshitzer et al. (1980). The axial gauge is interesting since it may be shown (most easily by the path-integral method - see Itzykson & Zuber (1980), Section 12-2-2) that ghosts are not needed even in the non-Abelian case.

5 Quantisation of vector fields: II Massive

The quantum theory of massive vector fields might be expected to be considerably easier than that of massless vector fields, since all the troubles attendant on unphysical degrees of freedom, which follow from a covariant approach to a vector field theory possessing a gauge-invariance, are presumably absent in a massive theory which, as we pointed out in Section 3.2, should have no such invariance. While this is perfectly true for the conventional massive vector field (Section 5.1 below), it turns out to be also irrelevant for, as we briefly indicated in Chapter 1, theories involving such fields are non-renormalisable - i.e. not calculable beyond lowest-order perturbation theory. We shall say more about this in Section 5.2. However, it is a remarkable fact that there exists one special class of massive vector field theories which is renormalisable - that in which there *is* a gauge symmetry present, but of the 'hidden' variety. In this case, as we shall see, the gauge fields do acquire mass, but renormalisability is not prejudiced. The GSW theory (Chapter 7) is of just this type. It might be expected, then, that the theory of this particular sort of massive vector field (the *only* acceptable possibility, presumably) would resemble quite closely the massless case, which is why we discussed the latter first. We shall approach the massive theory from this point of view, leading up to the discussion of hidden gauge symmetry in Chapter 6.

5.1 Massive vector fields (1)

A plausible extension, to the massive case, of the classical equations (4.9) for the free massless vector field, would be to replace $\square$ by $\square + M^2$ (where M is the mass of the vector field), obtaining the wave equation

$$(\square + M^2)B^\mu - \partial^\mu \partial_\nu B^\nu = 0. \tag{5.1}$$

We shall use B^μ for the massive analogue of the $U(1)$ field. To quantise, we ask for the Lagrangian which gives (5.1). A possibility is

$$\mathcal{L}_B = -\tfrac{1}{4}F_{B\mu\nu}F_B{}^{\mu\nu} + \tfrac{1}{2}M^2 B_\mu B^\mu, \tag{5.2}$$

where

$$F_{B\mu\nu} = \partial_\mu B_\nu - \partial_\nu B_\mu. \tag{5.3}$$

We note the mass term, appearing with the opposite overall sign as compared with the scalar case (2.6): as usual, the $\mu = 0$ components of B_μ appear with the wrong sign, while the spatial components have the same sign as in (2.6).

There are two fundamental points to be made about (5.1) and (5.2). Firstly, we repeat that, with the mass term, there is now no gauge freedom of the sort

$$B^\mu(x) \to B^\mu(x) + \partial^\mu \chi(x) \tag{5.4}$$

in either (5.1) or (5.2). Secondly, taking the divergence of (5.1), we find that the condition

$$\partial \cdot B = 0 \tag{5.5}$$

emerges as a consequence of the field equation (it would also be true were B^μ to be coupled to a conserved current), and does not have to be imposed. Thus the four components of B^μ are correctly and 'automatically' reduced to three by (5.5). We suspect that the canonical quantisation of such a theory should be straightforward.

In the massless case, we had trouble with the inverse of the wave operator $\Box g^{\mu\nu} - \partial^\mu \partial^\nu$, and with the inconsistency between the condition $\partial \cdot A = 0$ and the canonical quantisation conditions. From (5.1), the propagator for the massive case should be (i-times) the inverse of $[(-k^2 + M^2)g^{\mu\nu} + k^\mu k^\nu]$. It is easy to verify that this *has* an inverse, which is equal to

$$[-g^{\mu\nu} + k^\mu k^\nu/M^2]/(k^2 - M^2). \tag{5.6}$$

Further, it can be shown (Bogoliubov & Shirkov 1980) that the constraint (5.5) is consistent with the canonical commutation relation, and with the equations of motion, in the quantum field case. We note that the massless theory can *not* be recovered from (5.6) simply by letting $M \to 0$. That this is a singular limit should not be too surprising, perhaps; after all, the free massive field does have three independent polarisation states, whereas the massless one had only two. Such a difference in the number of degrees of freedom is hardly likely to be revealed in a smooth variation of $M \to 0$. In fact, however, the class of massive vector theories in which we shall really be interested is such that a smooth $M \to 0$ limit does exist! This is precisely because, in these theories, there is a hidden gauge symmetry.

In order to see what is wrong with the simple massive vector theory with propagator (5.6), we now briefly consider the question of renormalisability of such a theory, and the associated problem of the violation of unitarity bounds. Again, there will be some overlap with the fuller discussion of Aitchison & Hey (1982).

5.2 Diseases in the simple massive vector theory

Since the physical application of the massive vector theory will eventually be to the weak interactions, as described by the GSW theory, we may as well consider a weak interaction amplitude to illustrate the trouble with the simple theory of the preceding section. The illustration will provide a useful background to the later discussion of the GSW theory, in Chapter 7.

We consider, then, the process

$$\nu_\mu \bar{\nu}_\mu \to W^+ W^-, \tag{5.7}$$

where the charged W's are the massive vector particles, presumed to mediate the weak force. According to the theory, there is a μ-ν_μ-W coupling of the $V-A$ type, so that the lowest order (Born) amplitude for (5.7), shown in Fig. 5.1, is proportional to

$$\begin{aligned} M_{\lambda_1 \lambda_2} &= g^2 \epsilon_\mu{}^{-(\lambda_2)*}(k_2) \epsilon_\nu{}^{+(\lambda_1)*}(k_1) \bar{v}(p_2) \gamma^\mu (1-\gamma_5) \\ &\times \frac{\not{p}_1 - \not{k}_1 + m_\mu}{(p_1 - k_1)^2 - m_\mu{}^2} \gamma^\nu (1-\gamma_5) u(p_1) \end{aligned} \tag{5.8}$$

where $\epsilon^\pm$ are the polarisation vectors of the W's.

To calculate the cross-section we must form $|M|^2$ and sum over the three states of polarisation for each of the W's. To do this, we need the analogue of (4.68) for the two photon polarisation states. The present case is really easier, since we do not have to worry about gauge transformations (yet). In the W rest frame, the spin-1 condition

$$\partial \cdot W = 0 \tag{5.9}$$

implies (making a normal mode expansion analogous to (4.26)) that the three polarisation vectors have vanishing time component. We can choose three real orthogonal ones to be

$$\left.\begin{aligned} \epsilon_\mu{}^{(1)} &= (0, 1, 0, 0) \\ \epsilon_\mu{}^{(2)} &= (0, 0, 1, 0) \\ \epsilon_\mu{}^{(3)} &= (0, 0, 0, 1), \end{aligned}\right\} \tag{5.10}$$

just as in Section 4.1 (but here we do not have to think about the $\lambda = 0$ state!).

Fig. 5.1. μ^- exchange graph for $\nu_\mu \bar{\nu}_\mu \to W^+ W^-$.

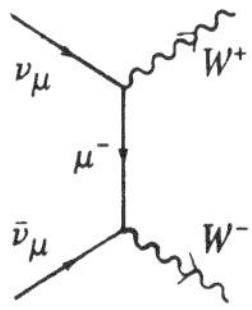

These satisfy

$$\epsilon_\mu{}^{(r)}\epsilon^{(s)\mu} = -\delta_{rs} \tag{5.11}$$

as well as the spin-1 condition $k \cdot \epsilon^{(r)} = 0$, in the rest frame. It is simple to check that, in this frame, we have

$$\sum_{\lambda=1}^{3} \epsilon_\mu{}^{(\lambda)}(k)\epsilon_\nu{}^{(\lambda)}(k) = -g_{\mu\nu} + k_\mu k_\nu/M^2. \tag{5.12}$$

We note the identity of this expression with the numerator in (5.6), the vector meson propagator. Boosting now to a frame such that $k^\mu = (k^0, 0, 0, |\mathbf{k}|)$, with $k^0 = (\mathbf{k}^2 + M^2)^{1/2}$, the $\lambda = 1, 2$ polarisation vectors are unchanged, but the longitudinal ($\lambda = 3$) one becomes

$$\epsilon_\mu{}^{(3)}(k) = \frac{1}{M}(|\mathbf{k}|, 0, 0, k^0), \tag{5.13}$$

which still satisfies $\epsilon \cdot k = 0$ and $\epsilon^2 = -1$.

Our interest will be in the high energy limit of this cross-section, and it is clear that it is the $k_\mu k_\nu/M^2$ which then dominates. We may write (5.13) as

$$\epsilon_\mu{}^{(3)}(k) = \frac{k^\mu}{M} + \frac{M}{k_0 + |\mathbf{k}|}(-1, \hat{\mathbf{k}}), \tag{5.14}$$

from which we can infer that it is the longitudinal polarisation state which is responsible for the leading high energy behaviour of the polarisation sum (5.12). This is an extremely important point. Retaining just these states in the cross-section, then, one finds

$$\sum_{\text{spins}} |M_{00}|^2 \propto \frac{E^4}{M^4}(1 - \cos^2\theta), \tag{5.15}$$

where E is the CM energy and θ the CM scattering angle. Putting all the requisite kinematic factors in, one obtains (Gastmans 1975)

$$\frac{d\sigma}{d\Omega} \propto \frac{E^2}{M^4}\sin^2\theta \tag{5.16}$$

and hence a total cross-section which rises with energy as E^2. However, the production of longitudinally polarised W's is actually a pure $J = 1$ process (see Gastmans 1975 for the partial wave decomposition), and the modulus of a single partial wave amplitude cannot exceed unity. It is apparent that this unitarity bound will be violated by (5.16), at some sufficiently large energy: if the proper factors are put in, one finds violation for $E \geqslant (6\pi/G_F)^{1/2}$, where G_F is Fermi's constant, related to the W-μ-ν coupling g by (see Chapter 7 below)

$$G_F/\sqrt{2} = g^2/8M_W{}^2. \tag{5.17}$$

The value of $(6\pi/G_F)^{1/2}$ is about 10^3 GeV.

At this stage it is instructive to go back to QED and see why, in terms of Born graphs for a similar process, the same trouble does not occur. Such a process is

$$e^+e^- \to \gamma\gamma, \tag{5.18}$$

which was already discussed, as it happens, in Section 4.4. There we saw that the k_μ-dependent parts of the photon polarisation sum (4.68) do not, in fact, contribute when the sum of the two relevant tree graphs (Figs 4.6 and 4.7) is considered. We know that these longitudinal parts must, indeed, be absent - by gauge-invariance - but the notion that a cancellation between different diagrams may be involved is a fruitful one.

Consider, for instance, the amplitude for

$$e^+e^- \to W^+W^-, \tag{5.19}$$

a kind of half-way house between (5.7) and (5.18). This has a lepton exchange tree graph, analogous to Fig 5.1, shown in Fig 5.2. As in the case of Fig 5.1, the amplitude will violate unitarity at high energies. But there is also a second tree graph contribution to (5.19), the one-photon intermediate state graph shown in Fig 5.3. This also violates unitarity at high energy, if at least one of the W's is longitudinal. Perhaps a cancellation could be arranged between these two tree graphs, as happened in the $e^+e^- \to \gamma\gamma$ case. Of course, in the present case the cancellation is much more exotic in character, as it would entail some relation between the *weak* process of Fig. 5.2, and the *electromagnetic* one of Fig. 5.3. Remarkably enough, just such a cancellation does occur in the GSW theory, in which the weak and electromagnetic interaction are related.

We may now return to the original example,

$$\nu_\mu \bar{\nu}_\mu \to W^+W^-. \tag{5.20}$$

In this case no one-photon Born graph like Fig. 5.3 is possible, and the

Fig. 5.2. $\bar{\nu}_e$ exchange graph for $e^+e^- \to W^+W^-$.

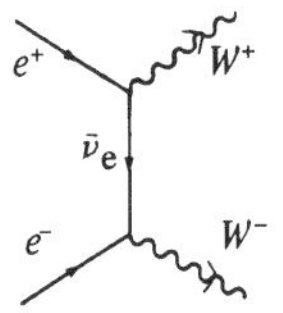

Fig. 5.3. One-γ intermediate state graph for $e^+e^- \to W^+W^-$.

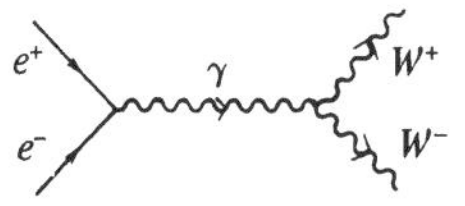

trouble seems incurable. Once again, however, the GSW theory will provide the way out: the existence of a *new* particle - the Z^0, a kind of heavy photon, which, however, interacts weakly (and is hence associated with a weak neutral current). A new Born graph, shown in Fig 5.4, occurs, whose high energy behaviour exactly cancels that of Fig 5.1.

This line of thought - that an acceptable massive vector theory can be arrived at by ensuring cancellation of 'bad' high energy behaviour of Born graphs - is capable of considerable extension. It can be shown (Llewellyn Smith 1974) to lead to models of the Georgi-Glashow (1972) type (in which new leptons are introduced) or of the GSW type (which introduce weak neutral currents and Higgs mesons - of which more in Chapter 7). Although unquestionably illuminating, this approach seems now to lack the depth provided by the more abstract gauge field concept - for the theories which the cancellation mechanism yields are precisely those involving a hidden gauge symmetry.

We add one word about renormalisability. The violations of unitarity we have been discussing here are rather different in principle from the ones (which lead to the idea of ghost contributions) discussed in Section 4.4. In the present case, all that has really been shown, it might be argued, is that perturbation theory fails at sufficiently high energies. The reason is, essentially, that the effective expansion parameter in the massive W theory is proportional to E/M, the tell-tale M^{-1} entering precisely via the longitudinal polarisation state (5.14). The ratio E/M can obviously become arbitrarily large. However, the real point is not the high energy growth of the Born graphs as such, but rather the disastrous effect this growth has on higher-order (loop) graphs, in which the Born graphs appear as components. For example, if we consider the box graph of Fig 5.5, which as usual involves an integral over a momentum k - say that of the W^+, we shall

Fig. 5.4. Z^0 intermediate state graph for $\nu_\mu \bar{\nu}_\mu \to W^+ W^-$.

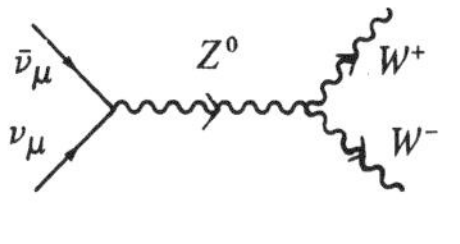

Fig. 5.5. Box graph contribution to $\nu_\mu \bar{\nu}_\mu \to \nu_\mu \bar{\nu}_\mu$, involving the tree graph of Fig. 5.1.

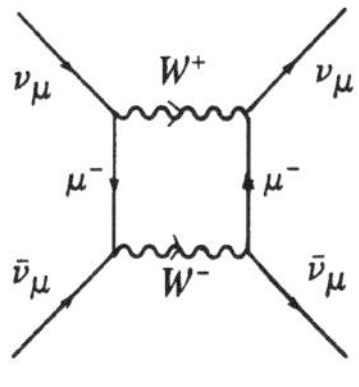

encounter an ultraviolet divergence as $k \to \infty$, associated precisely with the high energy rise of the Born graph of Fig 5.1, for a longitudinal W^+. This divergence is not renormalisable, and hence such a theory is not (at present) useful. In fact, the partial wave projections of Born amplitudes of renormalisable theories rise logarithmically with energy, while those of non-renormalisable theories seem invariably to grow with some power of the energy (cf. (5.15)). To go from Born (tree) graphs to loops is, however, not a straightforward step. There does not seem to be a rigorous proof, for example, that a necessary and sufficient criterion for renormalisability is that the partial wave projections of (the sums of all contributing) Born graphs grow at most logarithmically with energy. Nevertheless, such is the case for all known theories. Nor is the cancellation of non-renormalisable divergences always sufficient to ensure the validity of unitarity, in a perturbative sense: as Llewellyn Smith (1980) has remarked, one-loop divergences in some supersymmetric theories are cured by cancellation among tree graphs involving particles of different *spin*, yet clearly such cancellations cannot be invoked to save unitarity, for which only processes involving one definite set of external particles can be considered.

A particularly interesting case, in this connection, is that of 'massive QED' - i.e. the theory of a massive neutral vector meson coupled to a conserved current. One can check that the two graphs analogous to Figs 4.6 and 4.7 still cancel in the high energy limit, even though now $k_1{}^2 = k_2{}^2 \neq 0$. Indeed, this theory is renormalisable (Matthews 1949; Kamefuchi 1960; Salam 1960). In attempting to generalise this result to massive *charged* vector mesons, we arrive once again at the search for cancellation mechanisms among the Born graphs of the theory - only now we have linked the enterprise to renormalisability, rather than to perturbative unitarity. Rather than pursue this avenue further - a lucid and authoritative account exists already (Llewellyn Smith 1974) - we try instead to develop, for the massive case, the clue to the elimination of the longitudinal polarisation states which the massless theory suggests: namely, the improbable idea of a gauge symmetry for massive vector fields.

5.3 Massive vector fields (2)

The basic idea is to make the massive vector case as much like the successful QED case as possible. In particular, we should like to be able to 'gauge away' the offending $k^\mu k^\nu/M^2$ part of the propagator (5.6), or equivalently of the polarisation sum (5.12). We could do this in the massless case (4.44) by setting $\xi = 1$ (Feynman gauge). If we could introduce a similar gauge parameter in the massive case, we could hope to eliminate the $k^\mu k^\nu/M^2$ term, at least in one gauge (and then argue, by some kind of gauge-invariance, that it must in fact be harmless in all gauges). However, there seems to be no hope of such a thing, because the mass term $\frac{1}{2}M^2 B_\mu B^\mu$ in (5.2) destroys the $U(1)$ gauge-invariance.

All the same, suppose that, just for fun, we tried adding a gauge-fixing term to the massive spin-1 Lagrangian (5.2) - or, alternatively, a mass term to the gauge-fixed massless theory. This would produce a Lagrangian

$$\mathcal{L} = -\tfrac{1}{4}F_{A\mu\nu}F_A{}^{\mu\nu} - \frac{1}{2\xi}(\partial \cdot A)^2 + \tfrac{1}{2}M^2 A^\mu A_\mu, \tag{5.21}$$

in which

$$F_{A\mu\nu} = \partial_\mu A_\nu - \partial_\nu A_\mu .$$

The vector particle propagator would now be the inverse of

$$(-k^2 + M^2)g^{\mu\nu} + \left(1 - \frac{1}{\xi}\right)k^\mu k^\nu, \tag{5.22}$$

which is

$$\left[-g^{\mu\nu} + (1-\xi)\frac{k^\mu k^\nu}{k^2 - \xi M^2}\right]\Big/(k^2 - M^2). \tag{5.23}$$

As yet, (5.23) has no legitimacy whatever, but let us note the interesting features it exhibits:-

(i) good high energy behaviour, provided $\xi \neq \infty$, since the $k^\mu k^\nu$ term is divided by a k^2 factor;

(ii) the limit $M \to 0$ exists, and gives the gauge-fixed massless vector propagator (4.44);

(iii) $\xi \to \infty$ gives the ordinary massive vector propagator (5.6);

(iv) there is an extra pole at $k^2 = \xi M^2$.

Apart from (iv), (5.23) therefore looks very promising as a kind of 'hybrid' vector propagator with nice smooth limits to the conventional cases. But how can it be justified? The answer will lie in the 'hidden symmetry' - or 'spontaneous symmetry-breaking' - idea. An exact gauge symmetry will be spontaneously broken, and the gauge field will acquire mass. The mass will be proportional to the charge parameter of the theory, rather than being a totally independent parameter. The mass term can then be combined with other terms in the Lagrangian, and a transformation of the fields carried out so as to reduce the Lagrangian to manifestly gauge-invariant form.

The way this can come about may be illustrated by an alternative formulation of the massive vector theory, due originally to Stueckelberg (1938). We consider again $\mathcal{L}_B$ of (5.2), and introduce the decomposition

$$B^\mu \equiv A^\mu - \frac{1}{M}\partial^\mu \phi, \tag{5.24}$$

motivated by a desire to get at the $k^\mu k^\nu/M^2$ part of the propagator. Clearly, this decomposition is not unique: there is no change in B^μ if we make the changes

$$\left.\begin{aligned} A^\mu(x) &\to A^\mu(x) + \partial^\mu \chi(x) \\ \phi(x) &\to \phi(x) + M\chi(x). \end{aligned}\right\} \tag{5.25}$$

Clearly, this is some kind of local (gauge) transformation - at least the transformation for A^μ is; the interpretation of the ϕ transformation will appear in Section 5.4. If we rewrite $\mathcal{L}_B$ in terms of the A^μ and ϕ fields we obtain

$$\mathcal{L}_B = -\tfrac{1}{4}F_{A\mu\nu}F_A{}^{\mu\nu} + \tfrac{1}{2}M^2 A^\mu A_\mu - M\partial^\mu\phi A_\mu + \tfrac{1}{2}\partial_\mu\phi\partial^\mu\phi \tag{5.26}$$

which we interpret as follows. The last term in (5.26) describes a massless scalar field ϕ. The first two terms appear to describe a massive vector field A^μ. The third term looks like a ϕ–A coupling, with coupling constant proportional to the 'A' mass, M.

However, this interpretation cannot be quite right, because the number of degrees of freedom in (5.2) and in (5.26) would then not be equal: we started with *three* for the massive vector field B^μ in (5.2) and cannot end with *four* (three for A^μ and one for ϕ) in (5.26). Actually, it is easy to see that the 'A' part of (5.26) doesn't describe an ordinary massive spin-1 field, by trying to calculate the A-propagator. In doing so, we have to include the 'mixing term' $-M\partial^\mu\phi A_\mu$, which produces a coupling

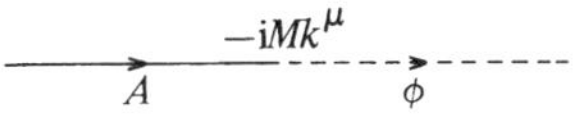

(5.27)

for a line of momentum k^μ. The complete propagator $S^{\mu\nu}$ is given by the series

(5.28)

summing which yields, formally,

$$S^{\mu\nu} = \frac{(g^{\mu\lambda} - k^\mu k^\lambda/M^2)}{(k^2 - M^2)}\,(g_\lambda{}^\nu - k^\nu k_\lambda/k^2)^{-1}. \tag{5.29}$$

But of course the required inverse in (5.29) fails to exist, as we saw in Section 4.1, so that all is not as it seems. Indeed, it is not surprising, now, that the A-propagator is undefined, since A^μ has the 'gauge' freedom (5.25), and hence we suspect that, as in QED, we shall have to fix the gauge of A^μ before we get a decent A-propagator.

A particularly clever way to do this was suggested by 't Hooft (1971*b*) (actually for the non-Abelian case - see Section 6.10). His suggestion for the present case is the addition of the gauge-fixing term $(-1/2\xi)(\partial_\mu A^\mu + M\xi\phi)^2$ to (5.26). This contains a term which in fact *cancels* the A-ϕ mixing term in

(5.26): $\mathcal{L}_B$ is now

$$\mathcal{L}'_B = -\tfrac{1}{4}F_{A\mu\nu}F_A{}^{\mu\nu} + \tfrac{1}{2}M^2 A_\mu A^\mu$$
$$-\frac{1}{2\xi}(\partial\cdot A)^2 + \tfrac{1}{2}\partial_\mu\phi\partial^\mu\phi - \tfrac{1}{2}\xi M^2\phi^2\,. \tag{5.30}$$

How are we to interpret (5.30)? The first three terms are exactly the ones in our peculiar proposal in (5.21)! But now we see that they come accompanied by the last two, describing a scalar field of mass $\sqrt{\xi}\,M$. The propagators are now

$$\left[-g^{\mu\nu} + \frac{(1-\xi)k^\mu k^\nu}{k^2 - \xi M^2}\right]\Big/(k^2 - M^2),\ \text{for the } A \text{ field} \tag{5.31}$$

$$(k^2 - \xi M^2)^{-1},\ \text{for the } \phi \text{ field.}$$

Let us pause to take stock. The point to hang on to is that we did after all start with a perfectly conventional free massive spin-1 theory. We have only *re-written* it, using five fields (A^μ, ϕ) instead of the four in B^μ. We can *not* attach physical significance to A^μ alone, or to ϕ alone - as is quite apparent from the presence of the unphysical 'gauge' parameter ξ in both their propagators. We can see this even more clearly by writing the A-propagator as

$$\frac{1}{k^2 - M^2}\left[-g^{\mu\nu} + \frac{k^\mu k^\nu}{M^2}\right] - \frac{k^\mu k^\nu}{M^2(k^2 - \xi M^2)}, \tag{5.32}$$

which shows that we have the usual massive propagator *plus* a ξ-dependent piece with a pole at $k^2 = \xi M^2$. This pole must actually be cancelled when we include the contributions from ϕ-lines. Indeed, by choosing $\xi \to \infty$ we can make it disappear altogether, along with the ϕ pole! In this 'gauge', therefore, the A^μ field by itself must provide a complete description of the physics. Going back to the transformations (5.25), we see that if we start with a given (A^μ, ϕ), and then choose the field $\chi(x)$ to be just $-\phi/M$, we end with $\phi' = 0$ and $A'^\mu = A^\mu - \partial^\mu\phi/M$, which is precisely the original B^μ field. The field ϕ has been 'gauged away'.

Returning to (5.30), we see that we can have our peculiar massive spin-1 propagator (with desirable $M \to 0$ and $k \to \infty$ limits) but at the expense of having introduced *also* an additional scalar field, $\phi(x)$, which transforms by (5.25), when A^μ suffers the indicated gauge transformation. However, its presence gets us out of our difficulty (iv), following (5.23). The choice of $\xi = 1$ in (5.31) gives the 'Feynman' gauge for the A field, and (as with any choice $\xi \neq \infty$) would *appear* to give us a renormalisable theory, because of the good high energy behaviour. The choice $\xi = 0$ gives a 'purely transverse' A-propagator, in the sense that $k_\mu[-g^{\mu\nu} + (k^\mu k^\nu/k^2)] = 0$.

In Section 4.2 we discussed the problem of negative norm states in the massless theory. In terms of the propagator (5.31) with $\xi = 1$, such contributions arise from the g^{00} part (the pole $k^2 = M^2$ will have a negative residue). In the case of QED the same thing occurs, but the g^{00} (scalar) contribution is cancelled there by the g^{33} (longitudinal) term, at the on-shell point $k^2 = 0$; as we saw in Section 4.1, the scalar and longitudinal photons cancel each other by virtue of condition (4.39), leaving only the positively-normed physical components at $k^2 = 0$. In (5.30), the g^{00} term will in fact be cancelled by the ϕ term (at $k^2 = M^2$ for $\xi = 1$).

We can summarise the position we have reached by saying that we are describing the free massive vector field B^μ by the five-component field (A^μ, ϕ), of which, of course, only three components are independent; the propagator for the A field is ξ-dependent, but so is that for the ϕ field, and the whole thing must be independent of ξ because the original B theory was. The general gauge $\xi \neq \infty$ shows a *possibility* of having a 'massive spin-1 field' - namely A^μ - behaving in a *renormalisable* way, *once we include the ϕ field along with special couplings and mass relations.* These relations will cause just the required cancellations among the various Born terms to render them harmless. Of course, there is no *real* question of renormalisation in this model, because we know all the time that it is equivalent to a *free* field. All the same, it is an instructional toy. If we could include some interaction, while keeping (5.26) as it is, we'd be home. This is just what the spontaneous symmetry breakdown of a local phase invariance will do. As a preliminary to that, we now show how (5.26) itself can be re-interpreted in local gauge field terms.

5.4 Re-interpretation of the A^μ - ϕ Lagrangian $\mathcal{L}'_B$

The purpose of this section is to examine in more detail the origin and interpretation of the transformations (5.25), which are allowed to the A^μ and ϕ fields. We have been tacitly regarding these transformations as some kind of gauge transformations, and this must now be explained. Indeed, we shall see that (5.25) is a type of local gauge-invariance (not of the electro-magnetic, *phase* invariance type, however); and we shall then, following Section 3.2, *reverse* the argument and *derive* the A-ϕ Lagrangian from a local gauge principle!

Our starting point† is the simplest imaginable field theory, that of a free massless scalar particle, with Lagrangian

$$\mathcal{L} = \tfrac{1}{2}\partial_\mu\phi\partial^\mu\phi. \qquad (5.33)$$

† The subsequent discussion owes much to the review article by Guralnik et al. (1968); the Lagrangian (5.29) is related to the model discussed by Boulware & Gilbert (1962).

This $\mathcal{L}$ is invariant under the *global* 'translation' (*not* phase change)

$$\phi(x) \to \phi(x) + \chi \quad \text{(global)}, \tag{5.34}$$

where χ is a constant (space-time independent). The current corresponding to this symmetry is

$$j^\mu = \partial^\mu \phi \tag{5.35}$$

and $\partial_\mu j^\mu = 0$ by virtue of the equation of motion for ϕ, $\Box\phi = 0$.

Suppose we now decide, following the philosophy of Section 3.2, to make the symmetry (5.34) a *local* one. We then require invariance of $\mathcal{L}$ under

$$\phi(x) \to \phi(x) + g\chi(x) \quad \text{(local)}, \tag{5.36}$$

where g is a constant introduced for later convenience. Clearly, (5.33) is *not* invariant under (5.36), but we can arrange for it to be so if we introduce a vector field A_μ with a certain well-known transformation property. Under (5.36),

$$\partial_\mu \phi \to \partial_\mu \phi + g\partial_\mu \chi, \tag{5.37}$$

so that if we introduce A_μ such that

$$A_\mu \to A_\mu + \partial_\mu \chi, \tag{5.38}$$

the combination

$$\partial_\mu \phi - gA_\mu \tag{5.39}$$

is *invariant* under (5.36) and (5.38). Thus the Lagrangian

$$\mathcal{L}_g = \tfrac{1}{2}(\partial_\mu \phi - gA_\mu)(\partial^\mu \phi - gA^\mu) - \tfrac{1}{4}F_{A\mu\nu}F_A{}^{\mu\nu} \tag{5.40}$$

implements the *local* transformation (5.36) in a 'minimal' way. We note from (5.38) that the dimension of the field χ is zero, and hence g in (5.36) has mass dimension 1. Setting $g = M$ we find that $\mathcal{L}_g$ is simply $\mathcal{L}_B$ of (5.26), and (5.36) and (5.38) are just (5.25). We also remark that the part of $\mathcal{L}_B$ which is linear in A^μ is $-MA_\mu j^\mu$, where j^μ is the current of the original global symmetry (5.34).

5.5 Hidden symmetry aspect

We can surround the above 'derivation' of the A_μ-ϕ Lagrangian with still further verbiage, which will actually be the language used in less simple cases. The original Lagrangian $\frac{1}{2}\partial_\mu \phi \partial^\mu \phi$ had the global invariance (5.34), but we should note that although this is a symmetry of the Lagrangian, it is *not* unitarily implemented in the space of states. We can see this as follows. If it were so implementable, we would be able to construct a unitary operator $U(x)$ such that (cf. 2.32)

$$\phi'(x) = \phi(x) + \chi = U(\chi)\phi(x)U^{-1}(\chi), \tag{5.41}$$

where

$$U = \exp(\mathrm{i}\chi R) \tag{5.42}$$

in terms of some Hermitian generator ('charge') R; R would be constructed from the a's and a^+'s of the field ϕ. The vacuum $|0\rangle$ of this theory would be defined by the condition that no ϕ quanta are present:

$$a(k)|0\rangle = 0 \tag{5.43}$$

as usual. We then expect that R acting on this empty state gives zero (the vacuum has zero eigenvalue of the charge) or, equivalently, 'the charge annihilates the vacuum', so that the vacuum is invariant under U:

$$(|0\rangle)' = U(\chi)|0\rangle = |0\rangle. \tag{5.44}$$

However, there is a contradiction between (5.41) and (5.44). For certainly,

$$\langle 0|\phi(x)|0\rangle = 0, \tag{5.45}$$

so that if

$$\phi'(x) = \phi(x) + \chi, \tag{5.46}$$

we find

$$\langle 0|\phi'(x)|0\rangle = \chi; \tag{5.47}$$

whereas from (5.41) and (5.44),

$$\begin{aligned} \langle 0|\phi'(x)|0\rangle &= \langle 0|U(\chi)\phi(x)U^\dagger(\chi)|0\rangle \\ &= \langle 0|\phi(x)|0\rangle = 0 \end{aligned} \tag{5.48}$$

(using the presumed zero eigenvalue of R), which contradicts (5.47). Thus the Lagrangian has a symmetry, but it is not representable in terms of the action of unitary operators in the physical space of states. In this situation, the symmetry is said to be 'hidden' or 'spontaneously broken'. We shall devote the next chapter to a detailed discussion of this phenomenon. We note here merely that the ϕ field associated with this peculiar symmetry is *massless* - if it had had a mass, the translation symmetry (5.34) would have been broken *explicitly*, not 'spontaneously'. The essential point, however, is that the massless ϕ field becomes combined with the massless gauge field A^μ in (5.39) to form a massive vector field (cf. the remarks at the end of Section 5.3). This is no isolated freak phenomenon, but possibly the simplest example illustrating the *generation of mass for vector particles in a spontaneously broken gauge theory.* Since the generation of the mass for the vector bosons in the Glashow-Salam-Weinberg theory of weak interactions comes about in precisely such a way, these concepts are now part of the essential 'tools of the trade'.

6 Symmetry in quantum field theory: II Hidden

We have advertised the fact that the GSW theory makes use of a hidden gauge-(local) invariance. Apart from that case, however, there is a further important application of the hidden symmetry idea in particle physics – the spontaneous breakdown of a global chiral symmetry (which does not involve the generation of vector boson masses). We shall therefore give a reasonably general discussion, including both the global and local cases. As we shall see, the physical consequences of hidden symmetry are quite different in the two cases. We shall discuss the global case first, but as a preliminary to both we begin with a general theorem.

6.1 The Fabri–Picasso theorem

Suppose that a given Lagrangian $\mathcal{L}$ is invariant under some one-parameter continuous global internal symmetry with a conserved Noether current j^μ, $\partial_\mu j^\mu = 0$. The associated charge is the Hermitian operator $Q = \int j^0 \, d^3\mathbf{x}$, and $\dot{Q} = 0$. We have hitherto assumed (though the nature of the assumption has been emphasised) that the transformations of this $U(1)$ group are representable on the Hilbert space $\mathcal{H}$ of physical states by unitary operators $U(\lambda) = e^{i\lambda Q}$ for arbitrary λ, with the vacuum invariant under U, so that $Q|0\rangle = 0$. Fabri & Picasso (1966) showed that there are actually just *two* possibilities:

(a) $Q|0\rangle = 0$, and $|0\rangle$ is an eigenstate of Q with eigenvalue 0;

or

(b) $Q|0\rangle$ does not exist in $\mathcal{H}$ (its norm is infinite).

The statement (b) is technically more true than the sometimes preferred statements '$Q|0\rangle \neq 0$' or '$Q|0\rangle = |0\rangle'$', although the latter does embody the useful physical idea of 'degenerate vacua' (see below). In case (a) we say that the symmetry is 'manifest', or that it is 'realised in the Wigner–Weyl way'; in (b) the symmetry is 'hidden', or 'spontaneously broken', and 'realised in the Nambu–Goldstone way'.

To prove this result, consider the vacuum matrix element $\langle 0|j_0(x)Q|0\rangle$. From translation invariance, implemented by the unitary operator $e^{-iP\cdot x}$ (where P^μ is the four-momentum operator) we obtain

$$\begin{aligned}&\langle 0|j_0(x)Q|0\rangle\\&\quad=\langle 0|\,e^{-iP\cdot x}j_0(0)\,e^{iP\cdot x}Q|0\rangle\\&\quad=\langle 0|\,e^{-iP\cdot x}j_0(0)\,Q\,e^{iP\cdot x}|0\rangle,\end{aligned}\tag{6.1}$$

where the last line follows from

$$[Q, P^\mu]=0,$$

since Q is an internal symmetry. Since the vacuum is a state of zero energy and momentum, we find

$$\langle 0|j_0(x)Q|0\rangle=\langle 0|j_0(0)Q|0\rangle,\tag{6.2}$$

which states that the matrix element we started from is in fact independent of x. We now consider the norm of $Q|0\rangle$:

$$\begin{aligned}\langle 0|QQ|0\rangle&=\int d^3\mathbf{x}\,\langle 0|j_0(x)Q|0\rangle\\&=\int d^3\mathbf{x}\,\langle 0|j_0(0)Q|0\rangle,\end{aligned}\tag{6.3}$$

which diverges unless $Q|0\rangle=0$. Thus either $Q|0\rangle$ has infinite norm, or the vacuum is annihilated by Q.

The foregoing has assumed the existence of a symmetry, and an associated conserved current. Remarkably enough, a theorem of Coleman (1966) proves a sort of 'converse' to the Fabri–Picasso theorem. We shall simply state Coleman's theorem here: if an operator

$$Q(t)=\int j_0(x)\,d^3\mathbf{x},$$

the space integral of a four-vector (*not assumed* to be conserved), annihilates the vacuum

$$Q(t)|0\rangle=0,$$

then in fact

$$\partial_\mu j^\mu=0$$

and the current is, after all, conserved; $Q(t)$ is then actually independent of t, and the symmetry is unitarily implementable by operators $U=\exp(i\lambda Q)$.

It is remarkable that it is the property of the vacuum, $|0\rangle$, which is apparently crucial in categorising how the symmetry is implemented.

6.2 The Goldstone theorem

We now want to investigate the consequences of possibility (b) of the Fabri-Picasso theorem – case (a) is the standard one, leading to the familiar phenomena of mass multiplets, etc., discussed in Chapter 2. An essential result, in case (b), was proved by Goldstone (1961); see also Goldstone, Salam & Weinberg (1962). Goldstone showed that when a continuous symmetry is hidden (case (b)) massless particles will necessarily be present in the theory. Whether these particles are actually observable or not depends, however, on whether the theory also contains gauge fields or not (i.e. on whether the symmetry is local, or not) – as we shall see in Section 6.4 and subsequent sections. We shall not prove Goldstone's theorem in full detail here, but give enough of a derivation to introduce the main ideas.

Suppose, then, that we have a Lagrangian $\mathcal{L}$ with a continuous symmetry generated by the charge Q, which is independent of time, and is the space integral of the $\mu = 0$ component of the Noether current:

$$Q = \int j_0(x)\, \mathrm{d}^3\mathbf{x}. \tag{6.4}$$

We consider the case in which the vacuum of this theory is not invariant i.e. is not annihilated by Q.

Suppose $\phi(y)$ is some field operator which is not invariant under the continuous symmetry in question, and consider the vacuum expectation value

$$\langle 0|[Q, \phi(y)]|0\rangle. \tag{6.5}$$

If Q were to annihilate $|0\rangle$, this would clearly vanish; we investigate the consequences of it *not* vanishing. Since ϕ is not invariant under Q, the commutator in (6.5) will give some other field, call it $\phi'(y)$; thus the hallmark of the hidden symmetry situation is the existence of some field (here $\phi'(y)$) with *non-vanishing vacuum expectation value.*

From (6.4), we can write (6.5) as

$$0 \neq \langle 0|\phi'(y)|0\rangle$$

$$= \langle 0|\left[\int \mathrm{d}^3\mathbf{x} j_0(x), \phi(y)\right]|0\rangle. \tag{6.6}$$

Since, by assumption, $\partial_\mu j^\mu = 0$, we have as usual

$$\frac{\partial}{\partial x^0}\int \mathrm{d}^3\mathbf{x} j_0(x) + \int \mathrm{d}^3\mathbf{x}\ \mathrm{div}\,\mathbf{j}(x) = 0, \tag{6.7}$$

whence

$$\frac{\partial}{\partial x^0}\int d^3\mathbf{x}\,\langle 0|[j_0(x),\phi(y)]|0\rangle$$
$$=-\int d^3\mathbf{x}\,\langle 0|[\operatorname{div}\mathbf{j}(x),\phi(y)]|0\rangle$$
$$=-\int d\mathbf{S}\cdot\langle 0|[\mathbf{j}(x),\phi(y)]|0\rangle. \tag{6.8}$$

If the surface integral vanishes in (6.8), (6.6) will therefore be independent of x_0. Leaving aside the 'if' for the moment, let us see where the independence of x_0 leads us. Inserting a complete set of states in (6.6), we obtain

$$0\neq\int d^3\mathbf{x}\sum_n\{\langle 0|j_0(x)|n\rangle\langle n|\phi(y)|0\rangle-\langle 0|\phi(y)|n\rangle\langle n|j_0(x)|0\rangle\}$$
$$=\int d^3\mathbf{x}\sum_n\{\langle 0|j_0(0)|n\rangle\langle n|\phi(y)|0\rangle\,e^{+ip_n\cdot x}$$
$$-\langle 0|\phi(y)|n\rangle\langle n|j_0(0)|0\rangle\,e^{-ip_n\cdot x}\}, \tag{6.9}$$

using translation invariance, with p_n the four-momentum eigenvalue of the state $|n\rangle$. Performing the spatial integral on the right-hand side we find

$$0\neq\sum_n\delta^3(\mathbf{p}_n)[\langle 0|j_0(0)|n\rangle\langle n|\phi(y)|0\rangle\,e^{ip_{n0}x_0}$$
$$-\langle 0|\phi(y)|n\rangle\langle n|j_0(0)|0\rangle\,e^{-ip_{n0}x_0}]. \tag{6.10}$$

But this expression is independent of x_0. *Massive* states $|n\rangle$ will produce explicit x_0-dependent factors $e^{\pm iM_nx_0}$ ($p_{n0}\to M_n$, as the δ-function constrains $\mathbf{p}_n=\mathbf{0}$), hence the matrix elements of j_0 between $|0\rangle$ and such a massive state must *vanish*, and such states contribute zero to (6.6). Equally, if we take $|n\rangle=|0\rangle$, (6.9) vanishes identically. But it has been assumed to be *not* zero. Hence *some* state or states must exist among $|n\rangle$ such that $\langle 0|j_0|n\rangle\neq 0$ and yet (6.9) is independent of x_0. The only possibility is states whose energy p_{n0} goes to *zero* as their three-momentum does (from $\delta^3(\mathbf{p}_n)$). Such states are, of course, *massless*. Thus the existence of a non-vanishing vacuum expectation value for a field, in a theory with a continuous symmetry, appears to lead inevitably to the necessity of having a massless particle, or particles, in the theory. This is the Goldstone result.

We can now interpret our little model of Sections 5.3 and 5.4 in these terms. The continuous global symmetry $\phi(x)\to\phi(x)+\chi$ *is* broken 'spontaneously', as we pointed out in Section 5.5, because there must exist a $\phi'(x)$ such that $\langle 0|\phi'(x)|0\rangle\neq 0$. There *is* a massless particle present – the ϕ-quantum itself. These two facts are now seen to be related. Also, the state of one ϕ-particle is connected to $|0\rangle$ by $j_0=\dot{\phi}$.

It is fair to say that there are fundamental unsolved questions involved in the generalisations of this result to arbitrary interacting quantum field theories. Most people talk as if the fields they were manipulating were classical fields; but there are questions of consistency raised in simply *postulating* (as is usually done) the existence of a certain non-vanishing vacuum expectation value. Guralnik et al. (1968) express reservations. In the trivial case of Section 5.5 ($\mathcal{L} = \frac{1}{2}\partial_\mu\phi\partial^\mu\phi$), we can exhibit the massless particle and prove the non-vanishing of a vacuum expectation value. But in the cases of interest to be discussed later, the hidden symmetry situation will be simply *postulated* via some $\langle 0|\phi|0\rangle \neq 0$.

6.3 The ferromagnet

The discussion so far has tended, it must be confessed, towards the formal. An illustration of the ideas in a more familiar (non-relativistic) context is helpful.

We have seen that the properties of the vacuum state are all-important. An essential aid to understanding hidden symmetry in quantum field theory is provided by Nambu's (1960) remarkable insight that the *vacuum* state of a quantum field theory is analogous to the *ground* state of an interacting many-body system. It is the state of lowest energy - the equilibrium state, given the kinetic and potential energies as specified in the Lagrangian. Now the ground state of a complicated system (for example, one involving interacting fields), may well have unsuspected properties - which may, indeed, be very hard to prove as following from the Lagrangian. But we can postulate (even if we cannot yet prove) properties of the quantum field theory vacuum $|0\rangle$ which are analogous to those of the ground states of many physically interesting many-body systems - such as ferromagnets, superfluids and superconductors, to name a few with which we shall actually be concerned.

Now it is generally the case, in quantum mechanics, that the ground state of any system described by a Hamiltonian is non-degenerate. Sometimes we may meet systems in which apparently more than one state has the same lowest energy eigenvalue. Yet in fact none of these states will be the true ground state: tunnelling will take place between the various degenerate states, and the true ground state will turn out to be a unique linear superposition of them. This is, in fact, the only possibility for systems of finite spatial extent. However, in the case of fields (extending presumably throughout all space), the Fabri-Picasso theorem shows that there is an alternative possibility, which is often described as involving a 'degenerate ground state' - a term we shall now elucidate. In case (i) of the theorem, the ground state is unique. For, suppose that several ground states $|0, a\rangle$, $|0, b\rangle, \ldots$ existed, with the symmetry unitarily implemented. Then one ground state will be related to another by

$$|0, a\rangle = e^{i\lambda Q}|0, b\rangle \tag{6.11}$$

for some λ. However, in case (i) the charge annihilates a ground state, and so all of them are really identical. In case (ii), on the other hand, we cannot write (6.11) - since $Q|0\rangle$ does not exist - and we do have the possibility of many degenerate ground states. In simple models (for instance, the ultra simple one $\mathcal{L} = \frac{1}{2}(\partial_\mu\phi)(\partial^\mu\phi)$, introduced earlier) one finds that these alternative ground states are all *orthogonal* to each other, in the infinite volume limit. Thus we have the picture emerging of many different inequivalent 'possible worlds', each one built on one of the possible orthogonal ground states.

These ideas can be illustrated by the physical example of the ferromagnet. In this case the symmetry in question is rotational invariance. Any conventional Hamiltonian of the Heisenberg exchange type ($J\Sigma \mathbf{S}_i \cdot \mathbf{S}_j$) is certainly rotationally invariant, yet the ground state of an actual magnet, below the Curie temperature T_C, has the spins all aligned in a *particular* direction (Fig. 6.1), and is clearly not rotationally invariant ('the ground state breaks the symmetry'). Indeed, there are infinitely many alternative ground states, below T_C, in which all the spins are lined up together in *different* directions (Fig. 6.2). These states are all orthogonal, in the infinite volume limit.

When we consider one particular ground state, we have - in one sense - a very symmetrical situation ('all spins lined up'). But in another sense it is very lop-sided, because one particular direction in space has been singled out. *All* the possible directions of alignment are equally likely ground state configurations; by picking just one, one 'hides' the rotational symmetry. Note also that the higher states, built by exciting spins from 'the' ground state so chosen, all share this lack of manifest rotational symmetry.

We may now ask why this 'hidden rotational symmetry' situation is accompanied by massless modes (as predicted by the Goldstone theorem). The modes

Fig. 6.1. Ferromagnetic ground state below T_C, all spins aligned.

Fig. 6.2. Alternative possible ground states.

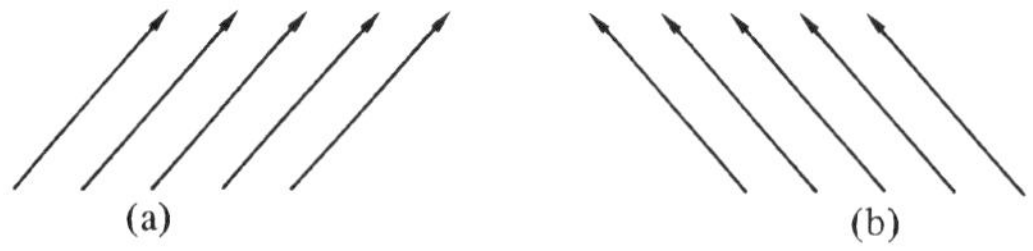

in this case are actually spin waves. Consider a spin wave of long wavelength λ (i.e. small k). Over regions of size $a \ll \lambda$ one has approximately a ground state with magnetisation in a particular direction (see Fig. 6.3). But corresponding to the existence of this excitation, the spins will rotate from region to region, with a characteristic spatial period of order λ. Now *it requires very little energy to bring about this excitation provided the spin-spin forces are of finite range.* In this case we can always find a large enough λ (small enough k) such that the effort to turn the spins over a scale λ is essentially zero - i.e. the energy required to excite the mode $\to 0$ as $k \to 0$. The spin-wave dispersion relation is actually $\omega \propto k^2$, which is like that of a massless particle. On the other hand, the presence of *long-range* forces will be such as to cause the surface integral in (6.8) *not* to vanish, and there will be no necessity for massless bosons: it will cost energy to rotate even widely-separated spins with respect to each other. This is a most important point, to which we shall return in subsequent sections.

In general, the massless modes may be thought of as generated by the operators which *would* rotate one vacuum into the other (if such operators existed!); after all, such a rotation should cost no energy. More precisely, we saw in the discussion of the Goldstone theorem that the massless modes were in the state $|n\rangle$ connected to the vacuum $|0\rangle$ by $j_0(0)$: $\langle 0|j_0(0)|n\rangle \neq 0$.

The ferromagnet gives us one further bit of useful physical insight. The Goldstone (massless) excitations are *space-time-dependent oscillations in the order parameter (field) whose non-vanishing ground state expectation value in the global theory defines the hidden symmetry situation.* Consider the ferromagnet, viewed 'globally'. The *total* spin vector of the magnet has a non-vanishing expectation value in the ground state $\langle 0|\mathbf{S}|0\rangle \neq 0$, and this defines the hidden symmetry. Going to a 'local' view, we consider $\langle 0|\mathbf{S}_i|0\rangle$ where $\mathbf{S}_i$ is the spin vector at lattice state i. In the (four-dimensional) continuum limit $\mathbf{S}_i \to \mathbf{S}(x)$, the field which is the local order parameter. It is the quantised oscillations in this field (spin waves) which are the Goldstone modes, such that $\omega \to 0$ as $k \to 0$. This idea is going to be important when we consider gauge fields, which *also* have a 'local' significance.

Fig. 6.3. Long wavelength excitation, involving rotation of neighbouring spins.

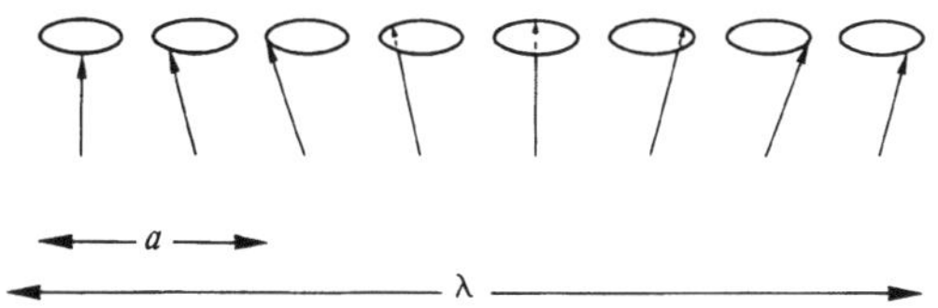

As a first introduction to the way the theorem gets crucially modified when gauge fields - or, alternatively, long-range forces - are present, we consider next an instructive non-relativistic model due to Kibble (1966).

6.4 The Kibble model

The model has the second-quantised Hamiltonian

$$H = \frac{1}{2}\int d^3\mathbf{x}\,[\pi^2(\mathbf{x},t) + (\nabla\phi(\mathbf{x},t))^2]$$

$$+ \frac{1}{2}\iint d^3\mathbf{x}\,d^3\mathbf{y}\,\pi(\mathbf{x},t)V(|\mathbf{x}-\mathbf{y}|)\pi(\mathbf{y},t), \tag{6.12}$$

where, with $\pi = \dot{\phi}$,

$$[\phi(\mathbf{x},t), \pi(\mathbf{y},t)] = i\delta^3(\mathbf{x}-\mathbf{y}). \tag{6.13}$$

This H is clearly invariant under

$$\phi(x) \to \phi(x) + \lambda \tag{6.14}$$

and is hence a kind of analogue of our model in Section 5.4. In particular, $\langle 0|\phi|0\rangle$ is *not* invariant under (6.14). In momentum space, quantising in a box of volume Ω,

$$H = \frac{1}{2}\sum_{\mathbf{k}} \{\mathbf{k}^2\phi_{\mathbf{k}}^*\phi_{\mathbf{k}} + [1 + V_{\mathbf{k}}]\pi_{\mathbf{k}}^*\pi_{\mathbf{k}}\}, \tag{6.15}$$

which is diagonal and therefore exactly soluble: clearly,

$$\dot{\pi}_{\mathbf{k}} = -\mathbf{k}^2\phi_{\mathbf{k}}^* \tag{6.16}$$

$$\dot{\phi}_{\mathbf{k}} = [1 + V_{\mathbf{k}}]\pi_{\mathbf{k}}^* \tag{6.17}$$

and so

$$\ddot{\phi}_{\mathbf{k}} = -\mathbf{k}^2[1 + V_{\mathbf{k}}]\phi_{\mathbf{k}} \tag{6.18}$$

and we find the dispersion relation

$$\omega^2 = \mathbf{k}^2(1 + V_{\mathbf{k}}) \tag{6.19}$$

for the ϕ modes. Now consider the behaviour as $k = |\mathbf{k}| \to 0$. Provided $k^2 V_{\mathbf{k}} \to 0$ as $k \to 0$, we will have $\omega \to 0$ as $k \to 0$; in fact $\omega^2 = k^2$ for small k, and the excitation is massless, $E^2 \propto k^2$. On the other hand, for a *Coulomb* potential of the form $V(\mathbf{x}) = g^2/|\mathbf{x}|$, the Fourier transform $V_k \sim g^2/k^2$. From (6.19) it is apparent that $V_{\mathbf{k}}$ is dimensionless, whence g is a *mass*, M. In this case $\omega^2 \to M^2$ as $k \to 0$, and the excitation is massive!

From this very simple example we can already draw the essential conclusions we shall need:

(a) a spontaneously broken symmetry such that $\langle 0|\phi|0\rangle \neq 0$ implies massless excitations; *but*

(b) this is *not* universally true without exception, but rather depends on the forces being of 'short enough range' – in this model the 'long-range' Coulomb force rendered the excitations massive.

We now turn to a final non-relativistic model, which will be very closely analogous to a quantum field theory case of interest.

6.5 The Bogoliubov superfluid

Consider the second-quantised many-body Hamiltonian

$$H = \frac{1}{2m}\int d^3\mathbf{x}(\nabla\psi)^\dagger \cdot \nabla\psi$$
$$+\frac{1}{2}\iint d^3\mathbf{x}\, d^3\mathbf{y}\, v(|\mathbf{x}-\mathbf{y}|)\psi^\dagger(\mathbf{x})\psi^\dagger(\mathbf{y})\psi(\mathbf{x})\psi(\mathbf{y})$$
$$-\mu\int d^3\mathbf{x}\, \psi^\dagger(\mathbf{x})\psi(\mathbf{x}), \tag{6.20}$$

where $\psi^\dagger(\mathbf{x})$ creates a boson at position $\mathbf{x}$. This H describes identical bosons† interacting via a potential v. The ordinary normal mode expansion for the operator $\psi(\mathbf{x})$ (in volume Ω)

$$\psi(\mathbf{x}) = \frac{1}{\sqrt{\Omega}}\sum_{\mathbf{k}} a_{\mathbf{k}}\, e^{i\mathbf{k}\cdot\mathbf{x}} \tag{6.21}$$

provides us with the standard particle interpretation in the non-interacting case, namely, that the ordinary ground state $|0\rangle$ will be the one annihilated by the $a_{\mathbf{k}}$ quanta, $a_{\mathbf{k}}|0\rangle = 0$. However, *below the superfluid transition temperature* T_B we know that in the limit as $v \to 0$ the true physical ground state is the zero-momentum state with *macroscopic occupation* – the 'bose condensate'. This means that below T_B the lowest energy level (ground state) of the non-interacting system is occupied by a significant fraction of the total number of the particles, N. More quantitatively, it is a standard result of statistical mechanics (see, for example, Tilley & Tilley (1974), Section 1.3) that for $T \leqslant T_B$ the ratio between the number $N_0(T)$ of particles in the ground state, and N, is (to a very good approximation)

$$\frac{N_0(T)}{N} = 1 - (T/T_B)^{3/2}, \tag{6.22}$$

† We have included a chemical potential term in (6.20); the restriction to a fixed number of particles, N, can be imposed to determine μ.

where

$$T_B \propto (N/\Omega)^{2/3}. \tag{6.23}$$

This feature necessitates a re-definition of the superfluid ground state in terms of the $a_{\mathbf{k}}$ operators, as we now discuss.

Bogoliubov (1947) argued that as $N \to \infty$ and $\Omega \to \infty$ such that $\rho = N/\Omega$ stayed finite, the effect of a *small* v (even at zero temperature) would be to induce a few excitations of particles to non-zero-momentum states, leaving the zero-momentum ground state still macroscopically occupied. In this case, we cannot have $a_0|0\rangle = 0$ for the ground state $|0\rangle$, or else clearly $\langle 0|a_0^\dagger a_0|0\rangle = 0$, rather than a number of order N. In fact, Bogoliubov argued, we may effectively replace both a_0 and $a_0^\dagger$ by the C-number value $N_0^{1/2}$ (roughly speaking because we may neglect the '1' on the right-hand side of $[a_0, a_0^\dagger] = 1$ by comparison with the $\sim N$ of $a_0^\dagger a_0$ - this can be made rigorous: see Guralnik et al. (1968)). If this replacement is made, and only terms of order N_0^2 and N_0 retained, the resulting approximate Hamiltonian can be exactly diagonalised in terms of the Bogoliubov *quasiparticle operators*

$$\alpha_{\mathbf{k}} = f_{\mathbf{k}} a_{\mathbf{k}} + g_{\mathbf{k}} a_{-\mathbf{k}}^\dagger \tag{6.24}$$

(and the corresponding $\alpha_{\mathbf{k}}^\dagger$) where the C-number functions $f_{\mathbf{k}}$ and $g_{\mathbf{k}}$ are real, spherically symmetric, and satisfy

$$f_{\mathbf{k}}^2 - g_{\mathbf{k}}^2 = 1. \tag{6.25}$$

Condition (6.25) ensures that the $\alpha_{\mathbf{k}}$, $\alpha_{\mathbf{k}}^\dagger$ operators obey the usual boson commutation relations, and so the transformation from the a's to the α's is canonical. The formalism is reviewed in Fetter & Walecka (1971), Chapter 10, for example.

For our purpose, the important point is the definition of the superfluid ground state. It is actually defined in terms of the quasiparticle operators, since they exactly diagonalise the (approximate) Hamiltonian. In fact, the 'Bogoliubov' ground state is the state $|0\rangle_B$ such that

$$\alpha_{\mathbf{k}}|0\rangle_B = 0, \text{ all } \mathbf{k} \neq \mathbf{0}; \tag{6.26}$$

i.e. the state with no non-zero-momentum quasiparticles in it. This is a complicated state in terms of the original $a_{\mathbf{k}}$ and $a_{\mathbf{k}}^\dagger$ operators. Indeed, the difficulty in finding the true ground state, and the brilliance of Bogoliubov's solution to that problem, sufficiently illustrate the point made in Section 6.3 about the possible complexity of the ground state in an interacting many-body system.

We now return to the question of the appropriate form of the normal mode expansion which should be adopted to deal with the special nature of this ground state. We have seen that the a_0, $a_0^\dagger$ operators are treated as C-numbers, so let us write

$$\psi(\mathbf{x}) = e^{i\alpha}\sqrt{\rho_0} + \chi(\mathbf{x}), \tag{6.27}$$

where α is an arbitrary phase, and ρ_0 is the number density of the zero-momentum state. We expand $\chi(\mathbf{x})$ as

$$\chi(\mathbf{x}) = \frac{1}{\sqrt{\Omega}} \sum_{\mathbf{k} \neq 0} a_{\mathbf{k}} \, e^{i\mathbf{k}\cdot\mathbf{x}}. \tag{6.28}$$

Then, since (inverting (6.24))

$$a_{\mathbf{k}} = f_{\mathbf{k}} \alpha_{\mathbf{k}} - g_{\mathbf{k}} \alpha_{-\mathbf{k}}{}^{\dagger}, \tag{6.29}$$

we find, using (6.26), that $\psi(\mathbf{x})$ has a non-zero expectation value in the ground state $|0\rangle_{\mathrm{B}}$:

$$_{\mathrm{B}}\langle 0|\psi(\mathbf{x})|0\rangle_{\mathrm{B}} = e^{i\alpha}\sqrt{\rho_0}. \tag{6.30}$$

This is the characteristic signal of a hidden symmetry, as we discussed in Section 6.2. We therefore expect massless modes to be present.

Indeed, the Bogoliubov diagonalisation produces the dispersion relation

$$\omega^2(\mathbf{k}) = \frac{\mathbf{k}^2}{2m}\left(\frac{\mathbf{k}^2}{2m} + 2\rho_0 v(k)\right). \tag{6.31}$$

Once again, if $v(k)$ is less singular than k^{-2} as $k \to 0$, ω will tend to zero as $k \to 0$ and we will have massless 'phonon-like' modes (if $v \sim$ constant, $\omega \sim k$). This type of excitation is observed experimentally at temperatures below T_{B}. On the other hand, if $v \sim k^{-2}$, $\omega \sim$ constant as $k \to 0$, and the quanta act as if they were massive (there is a 'gap' in the frequency spectrum, in that ω starts at a non-zero value for $k \to 0$).

These 'phonons' are the Goldstone bosons of the superfluid. The obvious question to ask now is: *what* continuous symmetry is broken spontaneously here? The answer is that the original H, (6.20), was invariant under the global $U(1)$ symmetry

$$\psi(\mathbf{x}) \to \psi'(\mathbf{x}) = e^{-i\theta}\psi(\mathbf{x}), \tag{6.32}$$

the generator being the number operator

$$Q = \int \psi^{\dagger}\psi \, d^3\mathbf{x}. \tag{6.33}$$

However, in our ground state $|0\rangle_{\mathrm{B}}$, we have

$$_{\mathrm{B}}\langle 0|\psi(\mathbf{x})|0\rangle_{\mathrm{B}} = e^{i\alpha}\sqrt{\rho_0}, \tag{6.34}$$

so that if $\psi(\mathbf{x})$ is transformed by (6.32), we obviously get

$$_{\mathrm{B}}\langle 0|\psi'(\mathbf{x})|0\rangle_{\mathrm{B}} = e^{i(\alpha-\theta)}\sqrt{\rho_0} \tag{6.35}$$

and the ground state $|0\rangle_{\mathrm{B}}$ does *not* respect the symmetry.

There is another way of looking at it, which is only rigorous for a finite system, but is nevertheless suggestive. The $U(1)$ phase transformations are generated by Q, and the unitary operator transforming ψ is $U = e^{iQ\theta}$ such that

$$\psi' = e^{iQ\theta}\psi\, e^{-iQ\theta} = e^{-i\theta}\psi, \text{ using } [Q, \psi] = -\psi. \tag{6.36}$$

So we have

$$\begin{aligned} {}_B\langle 0|\psi'|0\rangle_B &= e^{-i\theta}\,{}_B\langle 0|\psi|0\rangle_B \\ &= {}_B\langle 0|U\psi U^{-1}|0\rangle_B \neq {}_B\langle 0|\psi|0\rangle_B \end{aligned} \tag{6.37}$$

so that, clearly, if $|0\rangle_B$ *were* invariant under U (that is, $U|0\rangle_B = |0\rangle_B$) we should have a contradiction. Hence the ground state is *not* invariant under the symmetry transformation (but note that as $\Omega \to \infty$, Q becomes undefined). Indeed, there are *infinitely many degenerate ground states:* if we start with one particular one, call it $|\alpha\rangle$, such that $\langle\alpha|\psi(\mathbf{x})|\alpha\rangle = e^{i\alpha}\sqrt{\rho_0}$, since $[U, H] = 0$ we can equally well consider the degenerate state $U^{-1}|\alpha\rangle$ as an alternative ground state. We have seen that $U^{-1}|\alpha\rangle$ is not the same as $|\alpha\rangle$ - in fact it is clear that $U^{-1}|\alpha\rangle = |\alpha - \theta\rangle$, so the $U(1)$ transformation on a given ground state takes it to another degenerate one, of which there are infinitely many since θ is a continuous parameter. There is nothing to choose between them all, but when we *pick* the one such that, say ${}_B\langle 0|\psi|0\rangle_B = \sqrt{\rho_0}$ (i.e. $\alpha = 0$), we have singled one out preferentially and the symmetry among all the different ground states is lost. The parameter α distinguishes the different ground states, and the Goldstone 'phonons' are actually (though we do not show this here) oscillations in the *local* field variable $\alpha(x)$ (cf. end of Section 6.3).

One additional point: In forming the Bogoliubov transformation we might have been worried that *particle number* is not being conserved (ψ and $\psi^\dagger$ combinations). The particle number operator is Q, which is the space integral of $\psi^\dagger\psi$, which in turn is the $\mu = 0$ component if the four-current density j_μ whose spatial components are $\mathbf{j} = 1/2Mi(\psi^\dagger\nabla\psi - (\nabla\psi)^\dagger \cdot \psi)$, the matter flow current. If we expand the ψ field as in (6.21), we'd find (since $\int\psi^\dagger\psi\, d^3\mathbf{x} < \infty$) that $j_r \sim 1/r^3$ for large r, which is enough to show that, as $\Omega \to \infty$,

$$\frac{dQ}{dt} = 0$$

from $\partial_\mu j^\mu = 0$ (surface terms $\to 0$). But, if instead we use the expansion (6.27) and (6.28) with $\int_\Omega \chi^\dagger\chi\, d^3\mathbf{x} < \infty$, we find that j_r contains a piece $\sim 1/r^{3/2}$ (from the product $\sqrt{\rho_0}\, e^{i\alpha}(\partial\chi/\partial r)$) which is *not* enough to yield $dQ/dt = 0$. We can say that *in the infinite volume limit* 'the condensate provides an infinite reservoir of particles with which the observable quasiparticles (excitations) can exchange particle number'.

Finally, we return to the small k behaviour of $v(k)$. If v were Coulombic, $v(k) \sim q^2/k^2$, and then

$$\omega \sim |q|\sqrt{(\rho_0/m)} \text{ for small } k$$

which is just the 'plasma frequency'. The quasiparticles in this case are plasmons.

This is one important element in the BCS theory of superconductivity. Bosons of charge $-2e$ are formed by binding electron pairs near the Fermi surface; in the presence of the positive ions Bose condensation of pairs occurs, but the excitation spectrum is dominated by the Coulomb force between the pairs and is therefore plasmon-like with $q = |2e|$. (What is more, the electron spectrum itself has a 'gap', in the sense that instead of the dispersion relation $E = [(\hbar^2 k^2)/2m]$, we find $E^2 = [(\hbar^2 k^2/2m)^2 + \Delta^2]$.) The pairs in turn screen the Coulomb force, so that it becomes effectively $e^{-r/a}/r$ - that is, of *finite* range. In other words, the infinite range Coulomb force, mediated by the massless photon, has become a finite range force mediated by a particle of mass $\sim\hbar/ac$. We return to the superconductor in Section 6.9 below.

An important moral of this phenomenon seems first to have been voiced by Anderson (1963). In particle physics we do seem to have, empirically, a situation where symmetries exist, but are broken, to some extent. If we try and interpret the 'breaking' in terms of a hidden symmetry, we get landed with massless scalar (or, sometimes, pseudoscalar) quanta, which are experimentally in short supply. On the other hand, we saw much earlier how the 'gauge principle', though very elegant, *also* seemed impractical because the gauge quanta had to be massless. But we have just learned that if we have a hidden symmetry in the presence of long-range forces the Goldstone massless quanta acquire a mass. Indeed, the long-range force in the above case was itself a gauge force, with originally massless quanta, and *these* acquired a mass too. We seem to have a marvellous hope that in the case of a hidden gauge symmetry the *two* masslessnesses 'cancel each other out'! This is indeed the case, as we shall see in detail in Sections 6.9 and 6.10. We first discuss two fully relativistic examples of hidden *global* symmetry, beginning with the original classic model of Goldstone (1961).

6.6 Hidden Abelian global symmetry: the Goldstone model

We consider the case of a two-component scalar field (cf. Section 2.3)

$$\phi^\dagger = \frac{1}{\sqrt{2}}(\phi_1 + \mathrm{i}\phi_2), \phi = \frac{1}{\sqrt{2}}(\phi_1 - \mathrm{i}\phi_2). \tag{6.38}$$

Eventually, we shall introduce a gauge field, and then the ϕ field will be interpreted as having a charge. The Goldstone model takes

$$\mathcal{L} = (\partial_\mu \phi)^\dagger (\partial^\mu \phi) - V(\phi), \tag{6.39}$$

where V (the classical potential) has the form (cf. (2.93) - except that here ϕ is not an $SU(2)$ doublet)

$$V = \tfrac{1}{2}\lambda^2 |\phi|^4 - \tfrac{1}{2}\mu^2 |\phi|^2. \tag{6.40}$$

Then the Hamiltonian density is

$$\mathcal{H} = |\dot{\phi}|^2 + |\nabla\phi|^2 + V, \tag{6.41}$$

in which the first two terms are clearly the total energy of the free field. We treat this as a *classical* field problem. For the hidden symmetry situation, we are interested in the *ground state* configuration of the field. It is clear that the absolute minimum of $\mathcal{H}$ is reached for

(i) $\phi = \text{constant}$, which reduces the kinetic terms to zero;
(ii) $\phi = \phi_0$, where ϕ_0 is the minimum of V.

We note (a) $\lambda^2 > 0$ or else $\mathcal{H}$ has no lower bound, (b) if $\mu^2 < 0$ then

$$-\tfrac{1}{2}\mu^2 |\phi|^2 \to \tfrac{1}{2}|\mu^2|\,|\phi|^2$$

and $\mathcal{L}$ contains $-\frac{1}{2}|\mu|^2|\phi|^2$, the ordinary mass term as in (2.93). The *particle spectrum* (going to the quantum case) of the theory with $\mu^2 < 0$ is then found by looking at the free parts of $\mathcal{L}$ (particles being defined, presumably, by asymptotic states). In the present case, the terms $(\partial_\mu\phi)^\dagger\partial^\mu\phi - \frac{1}{2}|\mu|^2|\phi|^2$ are straightforwardly interpreted as two particles of equal mass $(1/\sqrt{2})|\mu|$. In this case, for $\mu^2 < 0$, V has the form shown in Fig. 6.4, which has a *unique* minimum at $\phi = \phi_0 = 0$. In the quantum field theory, we expand ϕ about the absolute minimum as usual

$$\phi = \frac{1}{\sqrt{\Omega}} \sum_{\mathbf{k}} (a_{\mathbf{k}+}\, \mathrm{e}^{-ik\cdot x} + a_{\mathbf{k}-}{}^\dagger\, \mathrm{e}^{ik\cdot x}), \tag{6.42}$$

where $a_\pm{}^\dagger = (1/\sqrt{2})\,(a_1{}^\dagger \pm i a_2{}^\dagger)$, and the ground state (vacuum) $|0\rangle$ is defined by

$$a_+|0\rangle = a_-|0\rangle = 0 \tag{6.43}$$

and clearly $\langle 0|\phi|0\rangle = 0$.

Now consider the case $\mu^2 > 0$. Then V has the form which has launched more than a thousand papers, and is shown in Fig. 6.5. It has a local maximum at $\phi = 0$, and a minimum along the circle $\phi_1^2 + \phi_2^2 = \mu^2/\lambda^2$. Here μ^2 is *not* a mass term, and we shall see that we are in the hidden symmetry situation. In fact, the potential V of (6.40), with $\mu^2 > 0$, is of just the form postulated for the Helmholtz free

Fig. 6.4. The shape of V (e.g. (6.40)) as a function of the fields ϕ_1, ϕ_2, for $\mu^2 < 0$.

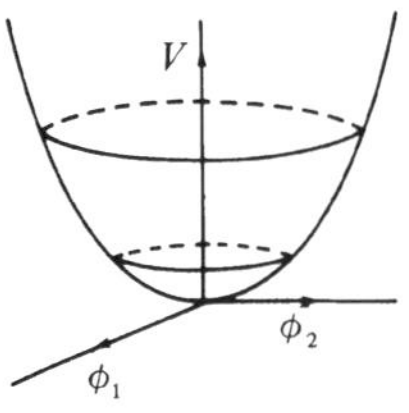

energy in Landau's (1937*a*, 1937*b*) 'mean field' theory of a second-order phase transition: for example, $|\phi|$ could be the *x-averaged* magnetisation in a ferromagnet. In that theory, μ^2 is a *temperature-dependent* quantity; above the ferromagnetic transition temperature T_C the effective potential has the form of Fig. 6.4 with only a minimum at the origin (no spontaneous magnetisation, $\mu^2 < 0$) while below it ($\mu^2 > 0$) the potential has the form of Fig. 6.5 and the magnetisation becomes non-zero. Then μ^2 is proportional to $(T_C - T)$ (>0 for $T < T_C$). For this reason, the word 'phase' is often used instead of 'situation' - as in 'the hidden symmetry phase'.

Returning to the quantum field theory case, this picture of V for $\mu^2 > 0$ *suggests* that the ground state (i.e. the field theory vacuum) should be the state (or, in fact, states) associated not with $\phi = 0$ but with $|\phi| = \text{constant} = (\mu^2/2\lambda^2)^{1/2}$. That is, the vacuum should be $|\omega\rangle_B$ (B for broken symmetry) such that

$$_B\langle\omega|\phi|\omega\rangle_B = \frac{1}{\sqrt{2}} e^{i\omega} |\mu|/|\lambda| = \frac{e^{i\omega}}{\sqrt{2}} f, \text{ say,} \tag{6.44}$$

where ω is some phase angle. It hardly needs to be said again that $_B\langle\omega|\phi|\omega\rangle_B$ is then not zero, and we are triggering the hidden symmetry phase, and expect to see massless bosons.

Before proceeding further it is as well to realise that this argument for the nature of the vacuum may be fine for classical fields, but its status is not so clear in the true quantum field theory case. A rigorous proof that the Lagrangian $\mathcal{L}$ of (6.39) with $\mu^2 > 0$ has indeed this $|0\rangle_B$ as the vacuum state, and that $_B\langle\omega|\phi|\omega\rangle_B \neq 0$ is consistent, seems not to be available. We shall accept it as an *assumption*.

Once again we enquire what the symmetry *is*, that is being spontaneously broken. $\mathcal{L}$ had the global $U(1)$ symmetry

$$\phi \to e^{-i\alpha}\phi, \text{ with constant } \alpha. \tag{6.45}$$

In case $\langle 0|\phi|0\rangle = 0$ ($\mu^2 < 0$), if $\phi' = e^{-i\alpha}\phi$ then clearly $\langle 0|\phi'|0\rangle = 0$ also, and (6.45) is respected by $|0\rangle$. But in case $\mu^2 > 0$, with (6.44) for $_B\langle\omega|\phi|\omega\rangle_B$, clearly

$$_B\langle\omega|\phi'|\omega\rangle_B = \frac{1}{\sqrt{2}} f\, e^{i(\omega-\alpha)} \neq {_B\langle\omega|\phi|\omega\rangle_B} \tag{6.46}$$

Fig. 6.5. The shape of V for $\mu^2 > 0$.

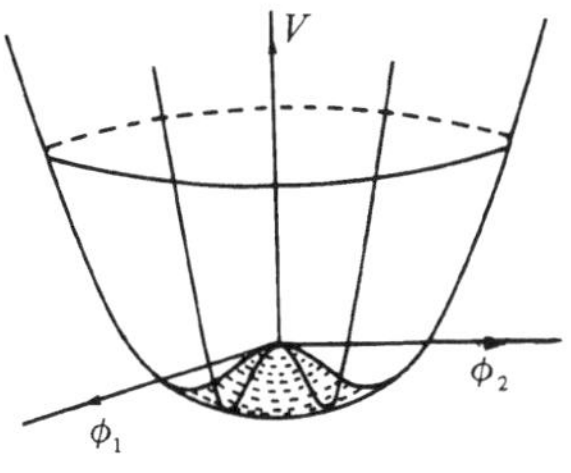

and '$|\omega\rangle_B$ is not invariant under the $U(1)$' (to speak loosely – as usual, the charge does not truly exist as $\Omega \to \infty$). Once more there is a single infinity of $|\omega\rangle_B$'s, distinguished by the phase parameter ω in (6.44). From Fig. 6.5, any point on the circle $|\phi|^2 = f^2/2$ will do: hence our labelling of the various vacua as $|\omega\rangle_B$. To build a theory we have to pick one of the vacua, thus destroying the $U(1)$ phase symmetry.

Having chosen one, say $\omega = 0$, the next task is to find the particle spectrum. To do this, we look at the free particle part of $\mathcal{L}$. But in case $\mu^2 > 0$, we cannot say that the $\frac{1}{2}\mu^2|\phi|^2$ piece represents a mass, because it has the wrong sign; and if we say that it is an interaction (with $(\partial_\mu\phi)^\dagger\partial^\mu\phi$ as the free part) we get simply a theory of massless quanta and a vacuum such that $\langle 0|\phi|0\rangle = 0$, if we expand ϕ as in (6.21), with $a|0\rangle = 0$. Instead we proceed by analogy with the *superfluid*, which was at least soluble (in some approximation), and so can serve as a guide to what may be sensible. Picking the value $\omega = 0$, and hence defining $|0\rangle_B$, we write

$$\phi_1 = f + \chi_1, \phi_2 = \chi_2, \tag{6.47}$$

since $\omega = 0$ corresponds to the minimum $\phi_1 = f, \phi_2 = 0$ on $\phi_1^2 + \phi_2^2 = f^2$, and we are looking for *small oscillations about this minimum*, whose quanta will be the physically interesting particles. Then we shall expect

$$_B\langle 0|\phi_1|0\rangle_B = f, \ _B\langle 0|\chi_1|0\rangle_B = 0 \tag{6.48}$$

and

$$_B\langle 0|\phi_2|0\rangle_B = {}_B\langle 0|\chi_2|0\rangle_B = 0. \tag{6.49}$$

Using (6.47), (6.39) becomes

$$\begin{aligned}\mathcal{L} = (\tfrac{1}{2}\partial_\mu\chi_1\partial^\mu\chi_1 - \tfrac{1}{2}\mu^2\chi_1^2) + \tfrac{1}{2}\partial_\mu\chi_2\partial^\mu\chi_2 \\ -\tfrac{1}{2}|\mu|\lambda\chi_1(\chi_1^2 + \chi_2^2) - \tfrac{1}{8}\lambda^2(\chi_1^2 + \chi_2^2)^2 + \mu^4/8\lambda^2.\end{aligned} \tag{6.50}$$

We disregard the constant, and interpret the rest of (6.50) as follows. The free-field parts are now easily interpretable as a part describing a field χ_1 of mass μ, and a part χ_2 which is a *massless* field. The global $U(1)$ symmetry has been spontaneously broken and we have one massless Goldstone boson. χ_1 corresponds to 'radial oscillations' about the chosen minimum of V, while χ_2 corresponds to zero-frequency motion 'around the circle', i.e. to modes having to do with *local oscillations in the phase angle which distinguished the different minima.*

Note that $\mathcal{L}(\chi_1, \chi_2)$ in (6.50) is *not* $U(1)$ symmetric in the sense of any naive transformation such as

$$\chi_1 \to \chi_1' = \cos\alpha\,\chi_1 - \sin\alpha\,\chi_2, \text{etc}. \tag{6.51}$$

Of course, it *is* invariant under the transformation

$$(f + \chi_1 + i\chi_2) \to (f + \chi_1 + i\chi_2)' = e^{-i\alpha}(f + \chi_1 + i\chi_2), \tag{6.52}$$

since this was the original global symmetry of $\mathcal{L}$. Thus $\mathcal{L}(\chi_1, \chi_2)$ is simple to interpret physically for $\mu^2 > 0$, but in it the symmetry is 'hidden'. Indeed, *it manifests itself in the peculiar relations between the various parameters of the theory*: (6.50) has three- and four-point couplings, and a mass μ, but there are only two independent parameters finally, λ and μ.

The *current* corresponding to the $U(1)$-invariance is

$$j_\mu = \mathrm{i}(\phi^\dagger \partial_\mu \phi - (\partial_\mu \phi)^\dagger \phi) = -(\phi_2 \partial_\mu \phi_1 - \phi_1 \partial_\mu \phi_2) \tag{6.53}$$

In terms of the χ_1, χ_2 fields, we find

$$-j_\mu = \chi_2 \partial_\mu \chi_1 - \chi_1 \partial_\mu \chi_2 - f \partial_\mu \chi_2, \tag{6.54}$$

so that

$$_{\mathrm{B}}\langle 0 | j_0(0) | n \rangle_{\mathrm{B}} \neq 0, \tag{6.55}$$

where $|n\rangle$ is a state with one χ_2 quantum of momentum P_n (cf. Section 6.2). In general, the current of the symmetry that is broken spontaneously will couple the vacuum to the Goldstone state, with a strength proportional to the non-vanishing vacuum value f.

Before concluding this section, it may be worth discussing the physics of the crucial step (6.47) somewhat further. $|0\rangle_{\mathrm{B}}$ is a very peculiar state when thought of in terms of the quanta of ϕ_1, but of course it is very simple in terms of the quanta of χ_1. In a sense we can regard the 'original' quanta of ϕ_1 as being like the material particles in a many-body system, while the most interesting things physically (*below the hidden symmetry phase transition point*) are not these, but the quanta of the χ_1 field. One may speculate that we are *literally* talking about a phase transition picture, exactly as in the superfluid or ferromagnet case. Possibly, in the very early stages of the universe, the temperature was above the critical (transition) value, and so we might imagine the manifest symmetry phase as occurring then, the present lop-sided state of affairs having arisen as the universe cooled. This would be represented phenomenologically by a temperature-dependent parameter μ^2.

We now generalise the foregoing to a non-Abelian global symmetry.

6.7 Hidden non-Abelian global symmetry

We can illustrate the essential features of this case and, moreover, prepare the way for the GSW theory, if we consider the Lagrangian (2.93), but with the sign of the μ^2 term reversed:

$$\mathcal{L} = (\partial_\mu \phi)^\dagger (\partial^\mu \phi) + \tfrac{1}{2}\mu^2 \phi^\dagger \phi - \tfrac{1}{2}\lambda^2 (\phi^\dagger \phi)^2 \tag{6.56}$$

with

$$\phi = \begin{pmatrix} \phi_{\mathrm{u}} \\ \phi_{\mathrm{d}} \end{pmatrix}$$

a doublet under $SU(2)$. The minimum of the potential now occurs at

$$(\phi^\dagger\phi)_0 = \mu^2/2\lambda^2 . \tag{6.57}$$

We need a generalisation of the '$\langle 0|\phi|0\rangle \neq 0$' condition. We recall that ϕ contains four real fields, and also that, as remarked in comment 6 at the end of Section 3.3, (6.56) is invariant under a global $U(1)$ phase group as well as under the global $SU(2)$, and so there are also four *generators* in play. The full continuous symmetry group G of (6.56) is $SU(2)\times U(1)$. In this more complicated non-Abelian situation, the following qualitatively new feature can arise: it may happen that the chosen condition of the form $\langle 0|\phi_i|0\rangle \neq 0$ (where i labels the component of the field transforming under G) is *invariant* under some subset of the transformations in G. This would effectively mean that *this* vacuum state *respected* that subset of symmetries, and thus we'd expect that there would be no massless Goldstone bosons corresponding to those particular charges (or currents). There would only be massless Goldstone bosons associated with each distinct transformation which did not leave $\langle 0|\phi_i|0\rangle$ invariant.

We reproduce the argument given in Taylor (1978). Suppose that $\langle 0|\phi_i|0\rangle = F_i$ where ϕ is a real r-component representation of G. In this representation the generators $Q^\alpha(\alpha = 1,\ldots,N)$ are represented by $\mathrm{i}v^\alpha$, where each v^α is a real antisymmetric $r\times r$ matrix. Then the vectors $v_{ij}{}^\alpha F_j(\alpha = 1,\ldots,N)$ span a subspace of dimension $n \leqslant r-1$, since clearly $F_i v_{ij}{}^\alpha F_j = 0$, so that all vectors in the subspace are orthogonal to $\mathbf{F}$. The potential V is a function of the ϕ_i, such that it is invariant under (cf. (2.95))

$$\phi_i \to \phi_i' = (\delta_{ij} + \epsilon^\alpha v_{ij}{}^\alpha)\phi_j . \tag{6.58}$$

Thus $\delta(V)=0$ gives

$$\frac{\partial V}{\partial\phi_i} v_{ij}{}^\alpha \phi_j = 0 . \tag{6.59}$$

We take $(\partial V/\partial\phi)_{\phi=\mathbf{F}}$ as our broken symmetry ground state condition. Differentiating (6.59) gives

$$\left(\frac{\partial^2 V}{\partial\phi_k\partial\phi_i}\right)_{\phi=\mathbf{F}} v_{ij}{}^\alpha F_j = 0 . \tag{6.60}$$

The factor in parentheses is the mass matrix for the ϕ fields, and from (6.60) we see that it annihilates all vectors in the n-dimensional subspace $v^\alpha F$. Thus there are n zero eigenvalues of the mass matrix or, in other words, n Goldstone bosons. The other $r-n$ scalars in ϕ_i will, in general, be massive.

An alternative way of looking at the situation is to ask whether some generator, or generators, or linear combination(s) of generators, of G is such that the corresponding matrix (or matrices) $c^\alpha v^\alpha$ annihilates F. Suppose that, for some particular

$\alpha, v^\alpha F = 0$; then under this particular transformation in G, $\delta\phi = [Q^\alpha, \phi] = v^\alpha \phi$ (cf. (2.81)) and so $\langle 0|\delta\phi|0\rangle = v^\alpha F = 0$. In other words, $\langle 0|\phi|0\rangle = \mathbf{F}$ is invariant under this particular part of G. The part of G that leaves $\langle 0|\phi|0\rangle = \mathbf{F}$ invariant is called the 'little group' of F in G. For these generators, $|0\rangle$ does not break the symmetry, and we expect no Goldstone phenomenon for them.

An example may make this clearer. Take ϕ to be an isovector in $SU(2)$, and take the generators in this three-dimensional representation to be

$$(v^\alpha)_{ij} = -\epsilon_{\alpha ij} \quad \text{(cf. (2.76))}. \tag{6.61}$$

Suppose that $\mathbf{F}$ is the vector $(1, 1, 1)$. The vectors $v^\alpha F$ are then

$$\begin{pmatrix} 0 \\ -1 \\ 1 \end{pmatrix}, \quad \begin{pmatrix} 1 \\ 0 \\ -1 \end{pmatrix}, \quad \begin{pmatrix} -1 \\ 1 \\ 0 \end{pmatrix}, \tag{6.62}$$

which clearly span a two-dimensional subspace ($n = 2, r = 3$). The condition

$$\langle 0|\phi|0\rangle = \mathbf{F} \tag{6.63}$$

is left invariant, it is clear geometrically, by rotations *about* the $\mathbf{n} = (1, 1, 1)$ direction. We can easily check, of course, that

$$n^\alpha v^\alpha = \begin{pmatrix} 0 & -1 & 1 \\ 1 & 0 & -1 \\ -1 & 1 & 0 \end{pmatrix} \tag{6.64}$$

gives zero when acting on $\mathbf{F}$. The combination (6.64) corresponds simply to a rotation about the direction of $\mathbf{n}$. The condition (6.63) would then lead to two of the components of ϕ being massless Goldstone particles; one component would be massive, corresponding to the unbroken symmetry of rotation about $\mathbf{n}$.

Consider then what condition we might impose on the isodoublet field ϕ we have elected (with one eye on the electroweak interaction application) to study in (6.56). By a choice of the global isospin 'coordinate system', and the phase freedom associated with the additional global $U(1)$, we can always reduce an arbitrary constant $\langle 0|\phi|0\rangle$, for isodoublet ϕ, to the form

$$\langle 0|\phi|0\rangle = \begin{pmatrix} 0 \\ f/\sqrt{2} \end{pmatrix}, \tag{6.65}$$

where, as before, $f = |\mu|/|\lambda|$. Pictorially, the situation can be depicted as in Fig. 6.6 (Maiani 1976): at each space-time point, there is a non-zero value for $\langle 0|\phi|0\rangle$, which is x-independent, and which by alignment of the 'internal' axes can be brought to the 'isospin pointing down' form of (6.65). We note that such a condition does not, in fact, break the full $SU(2)\times U(1)$ symmetry, because there is a combination of generators (in this two-dimensional representation) which annihilates the right-hand side of (6.65) (our previous 'F'), namely the combination $\frac{1}{2}(1+\tau_3)$. The other generators $\frac{1}{2}(1-\tau_3)$, $\tau_1/2$ and $\tau_2/2$ do not annihilate (6.65) and hence this bit of the full group *is* broken. We therefore expect to see one massive scalar field emerging when we expand about the minimum (6.65), and three massless fields. Furthermore, we shall see in Section 6.10 that when we make the symmetry *local* by including the non-Abelian gauge fields, we will end up with three massive vector particles (the massless Goldstones being 'swallowed' as their longitudinal components) and one massless vector particle – and of course this will be physically interpreted as the three vector bosons $W^{\pm}, Z^0$, and the usual photon, of the GSW model.

In order to study small oscillations about the minimum selected by (6.65) we need to set $\phi = \langle 0|\phi|0\rangle + \phi'$ and interpret the quanta of ϕ' as particles, as previously illustrated. Since $\langle 0|\phi|0\rangle$ has the isospinor structure

$$\begin{pmatrix} (0,0) \\ (f/\sqrt{2},0) \end{pmatrix}, \tag{6.66}$$

while ϕ involves two complex fields and has the general form

$$\begin{pmatrix} (\phi_1 + \mathrm{i}\phi_2)/\sqrt{2} \\ (\phi_3 + \mathrm{i}\phi_4)/\sqrt{2} \end{pmatrix}, \tag{6.67}$$

we shall write

$$\phi = \langle 0|\phi|0\rangle + \phi' \tag{6.68}$$

with

$$\phi' = \frac{1}{\sqrt{2}} \begin{pmatrix} \phi_2' + \mathrm{i}\phi_1' \\ \chi' - \mathrm{i}\phi_3' \end{pmatrix}. \tag{6.69}$$

(The point of the notation will become clearer when we consider the *local* version of this symmetry in Section 6.10.) Inserting (6.68), (6.66) and (6.69) into (6.56) and retaining only the quadratic terms, so as to investigate the mass spectrum, we find remarkably enough (but as expected from the general analysis) that the three fields ϕ' are *massless* while χ' has a mass μ. We may say (Maiani

Fig. 6.6. Heavy arrows represent the direction of $\langle 0|\phi|0\rangle$ in the 'internal' (isospin) space (*not* in space-time); this direction is the same for all x, t. The dotted lines represent the internal space axes.

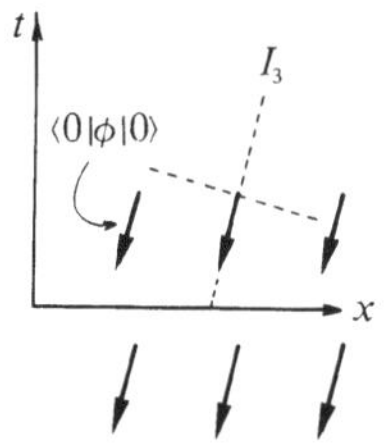

1976) that the non-zero vacuum expectation value of $\boldsymbol{\phi}$ is such that, at each point in space, a 'preferred direction' (the same for all x, in fact) is selected out, which causes wave propagation 'along it' (χ') to differ from that 'transverse' to it ($\boldsymbol{\phi}'$).

6.8 Hidden global chiral symmetry

Although apparently something of a diversion, the topic of this section does, in fact, have some relevance to gauge theories, as we shall see; it also provides a real example from particle physics of a global symmetry realised in the Nambu–Goldstone mode.

We begin by considering a simple model Lagrangian, namely,

$$\mathcal{L} = \bar{\psi}\mathrm{i}\not{\partial}\psi + \mathrm{i}g\bar{\psi}\boldsymbol{\tau}\gamma_5\psi\cdot\boldsymbol{\phi} + g\bar{\psi}\psi\sigma + \tfrac{1}{2}(\partial_\mu\boldsymbol{\phi})\cdot(\partial^\mu\boldsymbol{\phi}) + \tfrac{1}{2}(\partial_\mu\sigma)(\partial^\mu\sigma) - \tfrac{1}{2}\mu^2(\sigma^2+\boldsymbol{\phi}^2) - \tfrac{1}{4}\lambda^2(\sigma^2+\boldsymbol{\phi}^2)^2, \tag{6.70}$$

where the spinor field ψ is a *massless* doublet under a global $SU(2)$ group, the pseudoscalar field $\boldsymbol{\phi}$ is a triplet under $SU(2)$, and the scalar field σ is also an $SU(2)$ scalar. Though rather lengthy to write down, (6.70) is indeed quite familiar-looking by now: the $\psi-\boldsymbol{\phi}$ coupling (apart from the pseudoscalar aspect, indicated by the γ_5) was mentioned in section 2.3, while the $\boldsymbol{\phi}$ and σ pieces are standard free Langrangians for scalar particles (section 2.1) all of the same mass μ, together with an interaction term reminiscent of (6.56), and a simple $\psi-\sigma$ coupling.

What are the global symmetries of (6.70)? First of all, it is quite clearly invariant under the (infinitesimal) $SU(2)$ transformations

$$\left.\begin{aligned} \delta\psi &= -\mathrm{i}\boldsymbol{\epsilon}\cdot\boldsymbol{\tau}/2\ \psi \\ \delta\boldsymbol{\phi} &= \boldsymbol{\epsilon}\times\boldsymbol{\phi} \\ \delta\sigma &= 0. \end{aligned}\right\} \tag{6.71}$$

The associated current ($a = 1, 2, 3$)

$$j^{a\mu}(x) = (\boldsymbol{\phi}\times\partial^\mu\boldsymbol{\phi})^a + \tfrac{1}{2}\bar{\psi}\gamma^\mu\tau^a\psi \tag{6.72}$$

is conserved, and the charges

$$Q^a = \int j^{a0}(x)\,\mathrm{d}^3\mathrm{x} \tag{6.73}$$

are constants of the motion. However, (6.70) is also invariant under a further set of transformations, namely,

$$\left.\begin{aligned} \delta\psi &= -\mathrm{i}\boldsymbol{\eta}\cdot\boldsymbol{\tau}/2\ \gamma_5\psi \\ \delta\boldsymbol{\phi} &= \boldsymbol{\eta}\sigma \\ \delta\sigma &= -\boldsymbol{\eta}\cdot\boldsymbol{\phi}, \end{aligned}\right\} \tag{6.74}$$

where the infinitesimal parameter is now denoted by η. There is therefore a second set of conserved currents

$$j_5^{a\mu}(x) = \tfrac{1}{2}\bar{\psi}\gamma^\mu\gamma_5\tau^a\psi + (\sigma\partial^\mu\phi^a - \phi^a\partial^\mu\sigma) \qquad (6.75)$$

and corresponding charges Q_5^a.

We make two initial comments about (6.74). Firstly, we note that these transformations involve changes in *parity* ($\psi \to \gamma_5\psi$, pseudoscalar $\phi \to$ scalar σ). As a consequence, the current (6.75) is an *axial* vector, and the charges Q_5^a are pseudoscalars. Secondly, we can easily verify that these axial currents would *not* be conserved if we added a fermion mass term.

We may now consider the *algebra* generated by the charges Q^a, Q_5^a. We find the interesting results

$$[Q^a, Q^b] = \mathrm{i}\epsilon_{abc}Q^c \qquad (6.76)$$

$$[Q^a, Q_5^b] = \mathrm{i}\epsilon_{abc}Q_5^c \qquad (6.77)$$

$$[Q_5^a, Q_5^b] = \mathrm{i}\epsilon_{abc}Q^c. \qquad (6.78)$$

(6.76) says that the Q^a are the generators of an $SU(2)$; (6.77) says that the Q_5^a's transform as an isovector under this $SU(2)$; and (6.78) *closes* the algebra of the six charges. Clearly, the relations (6.76)-(6.78) constitute a more complicated system than the familiar set (6.76) alone: we may wonder how to set about finding matrix representations of this algebra. Fortunately, a great simplification occurs if we consider the combinations

$$Q_R^a = \tfrac{1}{2}(Q^a + Q_5^a) \qquad (6.79)$$

$$Q_L^a = \tfrac{1}{2}(Q^a - Q_5^a), \qquad (6.80)$$

for then we find

$$[Q_R^a, Q_R^b] = \mathrm{i}\epsilon_{abc}Q_R^c \qquad (6.81)$$

$$[Q_L^a, Q_L^b] = \mathrm{i}\epsilon_{abc}Q_L^c \qquad (6.82)$$

$$[Q_R^a, Q_L^b] = 0. \qquad (6.83)$$

(6.83) shows that the Q_R's and Q_L's commute with each other, while (6.81) and (6.82) show that they each generate separate $SU(2)$ algebras. Their combined algebra is therefore called $SU(2)_L \times SU(2)_R$. The representations are then labelled by the eigenvalues of the 'left-handed isospin' Q_L^2 and of the 'right-handed isospin' Q_R^2. The terms 'left' and 'right' originate from the factors $(1 \pm \gamma_5)/2$ in the fermion contribution, which are just the helicity projection operators for a massless particle. Such a symmetry, involving 'left' and 'right' charges, is called a *chiral* one. We note that under the parity operation $Q_L^a \leftrightarrow Q_R^a$.

Just as the algebra of $SU(2)$ is the same as that of the generators of rotations in a three-dimensional space, so the algebra of $SU(2) \times SU(2)$ turns out to be the same as that of the generators of rotations in a four-dimensional (Euclidean) space - here, of course, an 'internal' space involving the field components. In fact, this is perhaps seen more clearly in terms of the original charges Q^a, Q_5^a and transformations (6.71) and (6.74). Clearly, (6.71) are the familiar 'rotations' in isospace for the $t=\frac{1}{2}$ isodoublet ψ, and $t=1$ triplet $\boldsymbol{\phi}$. The transformations of $\boldsymbol{\phi}$ and σ in (6.74) are a kind of infinitesimal Euclidean 'Lorentz transformation' of the 'four-vector' $(\sigma, \boldsymbol{\phi})$, with $\boldsymbol{\eta}$ an infinitesimal 'velocity'. Just as real (Minkowskian) Lorentz transformations preserve $t^2 - \mathbf{x}^2$, so (6.74) preserves the Euclidean form $\sigma^2 + \boldsymbol{\phi}^2$. It is just this form which is present in the Lagrangian (6.70), thereby leading to the invariance under (6.74).

We now observe that when $\mu^2 < 0$ the σ-$\boldsymbol{\phi}$ potential terms in (6.70) produce a four-dimensional analogue of Fig. 6.5, such that there is a minimum of the potential at a non-zero field value:

$$\langle\sigma^2 + \boldsymbol{\phi}^2\rangle_0 = -\mu^2/\lambda^2. \tag{6.84}$$

(6.84) generalises the circular minimum of Fig. 6.5 to the surface of a sphere in four dimensions; there are thus many equivalent configurations of $\boldsymbol{\phi}$, σ satisfying (6.84). As in the $U(1)$ case, however, we may choose a particular ground state, say

$$\langle\boldsymbol{\phi}\rangle_0 = \mathbf{0}, \langle\sigma\rangle_0 = (-\mu^2/\lambda^2)^{1/2} \equiv v. \tag{6.85}$$

To get the physical interpretation of (6.84), we expand about the minimum (6.85), and write (as in (6.47))

$$\sigma = v + \sigma'; \tag{6.86}$$

$\boldsymbol{\phi}$ does not need to be translated, by the choice (6.85). Introducing (6.86) into (6.70) produces some remarkable results. $\mathcal{L}$ becomes (disregarding a constant term)

$$\begin{aligned}\mathcal{L} = {} & \bar{\psi}(\mathrm{i}\not{\partial} + gv)\psi + \mathrm{i}g\bar{\psi}\boldsymbol{\tau}\gamma_5\psi\cdot\boldsymbol{\phi} + g\bar{\psi}\psi\sigma' \\ & + \tfrac{1}{2}(\partial_\mu\boldsymbol{\phi})\cdot(\partial^\mu\boldsymbol{\phi}) + \tfrac{1}{2}(\partial_\mu\sigma')(\partial^\mu\sigma') + \mu^2\sigma'^2 - \lambda^2 v\sigma'(\sigma'^2 + \boldsymbol{\phi}^2) \\ & - \tfrac{1}{4}\lambda^2(\sigma'^2 + \boldsymbol{\phi}^2)^2.\end{aligned} \tag{6.87}$$

We see that

(i) the fermion has acquired a mass $-gv$, which is proportional to the vacuum value v;

(ii) the $\boldsymbol{\phi}$ field is massless - these are the *three* Goldstone bosons, corresponding to the three directions 'perpendicular' to the one selected out by the symmetry-breaking condition (6.85);

(iii) the σ' field is massive, with mass $(-2\mu^2)^{1/2}$, corresponding to the condition (6.85);

(iv) there are new trilinear σ'-$\boldsymbol{\phi}$ couplings, proportional to v.

An important physical application of these ideas is to the QCD Lagrangian of (3.66). The colour fields are assumed to bind the quarks into hadrons, which are colour singlets. For massless quarks, this Lagrangian has a chiral symmetry of the type considered above, which operates in the internal *flavour* space (u, d, s, . . .) of the quark fields. In this case, only the first of the transformations (6.71) and (6.74) - or their generalisations to other flavours - is involved. For definiteness, we limit discussion to the $SU(2)$ flavour group; the chiral symmetry is then $SU(2)_L \times SU(2)_R$. The associated vector and axial vector currents (which are related to, though not identical with, currents that enter the gauge theory of weak interactions, to be described in Chapter 7) will be conserved, and will obey commutation relations of the sort satisfied by the model currents (6.72) and (6.75), while the corresponding charges obey (6.76) - (6.78). The $SU(2)$ *vector* currents are interpreted as isospin currents, and the symmetry associated with their conservation is manifest in the hadronic isospin multiplets. What about the symmetry corresponding to the conservation of the axial vector currents? Even if the quark masses in (3.66) were zero, this could not be realised as a manifest hadronic symmetry, for the baryons would then have to be massless. Instead, this axial $SU(2)$ symmetry is interpreted as realised in the Nambu-Goldstone mode. In this case the bound state fermions can be massive, but we have to accept the presence of massless pseudoscalar particles, which the pseudoscalar charges connect to the vacuum. The idea is then to identify these particles with the pion triplet (in the $SU(2)$ case). The Lagrangian (6.87) itself is then interpreted as an effective Lagrangian, exhibiting the original $SU(2)_L \times SU(2)_R$ symmetry spontaneously broken into the isospin group $SU(2)$.

Certainly, the pion mass is not zero, but it may be considered as small, perhaps. Another apparent difficulty with the above ideas is that the axial vector current cannot be exactly conserved, or else - as was shown long ago by Taylor (1958) - the pion itself would not decay. Actually, these difficulties seem to be related. We can imagine a world in which the pion is massless, and the axial currents exactly conserved - the hidden realisation of an exact chiral symmetry. We then introduce a non-zero pion mass m_π, such that the divergence of the axial vector current is proportional to m_π^2. The chirally-symmetric limit is then $m_\pi^2 \to 0$. This is called 'Partially Conserved Axial Vector Current Theory' (PCAC).

These ideas may be illustrated in a model of the form (6.70) by the addition of a term

$$c\sigma(x) \tag{6.88}$$

in the original Lagrangian, where c is a constant (this is then precisely the σ-model of Gell-Mann & Levy (1960)). With this explicit symmetry-breaking condition, $\partial_\mu j_5^{a\mu}(x)$ is no longer zero, but is given immediately from (6.74) as

$$\partial_\mu j_5^{a\mu}(x) = -c\phi^a(x). \tag{6.89}$$

The addition of (6.88) selects out one unique minimum of the potential; it is at a non-zero value of the field σ, which we again call v, and at a zero value of ϕ. This minimum is at the point such that

$$c = v(\mu^2 + \lambda^2 v^2). \tag{6.90}$$

When the shift (6.86) is carried out now, one finds a similar result to (6.87), with, however, the addition of a ϕ mass term

$$-\tfrac{1}{2}(\mu^2 + \lambda^2 v^2)\phi^2 \tag{6.91}$$

(the σ' mass also changes; in imposing (6.90) as the minimum condition, we are ignoring higher-order corrections to this classical result). (6.89) and (6.90) together give

$$\partial_\mu j_5^{a\mu}(x) = -v m_\pi{}^2 \phi^a(x), \tag{6.92}$$

which indeed vanishes as $m_\pi{}^2 \to 0$ ($c \to 0$ is the symmetry limit). By comparing (6.92) with the matrix element for $\pi \to e\nu$ decay, one finds that $v = -f_\pi$, the π decay constant. The PCAC hypothesis is then

$$\partial_\mu j_5^{a\mu}(x) = f_\pi m_\pi{}^2 \phi^a(x), \tag{6.93}$$

independent of this particular model.

In quark terms, introduction of quark masses, as in (3.66), will explicitly break the chiral symmetry; nevertheless, the commutation relations among the quark currents and charges may well survive, as is postulated in the 'current (and charge) algebra' of Gell-Mann (1961). Isospin symmetry will remain good to the extent to which $m_\mathrm{u} \approx m_\mathrm{d}$. The pion mass itself is, on the above view, due to quark masses - the origin of which is an unsolved problem (see further Sections 7.4 and 7.5 below).

It would take us too far afield to pursue the consequences of this theory here (see, for example, the book by Adler & Dashen 1968). We may note, however, that there is good evidence for the approximate validity of hidden chiral symmetry. This evidence includes the Goldberger-Treiman (1958) relation; low energy theorems for πN scattering (Nambu & Lurié 1962; Adler 1965*a*, 1965*b*); the Adler-Weisberger relation† (Adler 1965*c*, Weisberger 1965) and the Callan-Treiman (1966) relation. These results seem to be verified experimentally to an accuracy of about 10%. The idea can be extended to strangeness-changing axial vector currents, and works to an accuracy of about 30%.

These empirical successes of the (hidden) chiral symmetry idea - and the additional successful prediction of the $\pi^0 \to \gamma\gamma$ decay rate (Section 8.3) - can be used to provide some *a posteriori* justification for the QCD Lagrangian

† Derivation of these results will be indicated in Section 8.2 below.

(3.66) (Llewellyn Smith 1980); whatever the fundamental fields are, the strong forces between them should be such as to preserve this symmetry.

We now return to the main line of our discussion, and take up the question of the generation of mass for gauge quanta.

6.9 Hidden local $U(1)$ symmetry: electromagnetic interactions in the Goldstone model (or, the Higgs model)

We have seen in Sections 6.4 and 6.5 how in non-relativistic theories the massless Goldstone excitations have a habit of getting heavy when electromagnetic forces are included. So now we look at this phenomenon in the relativistic case, using the Higgs model (Higgs 1964*a*, 1964*b*). This model is just the Goldstone one, but with the addition of an electromagnetic interaction. This interaction is introduced via the gauge principle (Section 3.2): the interaction will be *determined* by requiring that the

$$\text{global } U(1) \text{ invariance } \phi(x) \to \mathrm{e}^{-\mathrm{i}\alpha}\phi(x) \tag{6.94}$$

be replaced by a

$$\text{local } U(1) \text{ invariance } \phi(x) \to \mathrm{e}^{-\mathrm{i}e\chi(x)}\phi(x). \tag{6.95}$$

A charge e has been put in for later convenience. Then, as we discussed in Section 3.2, to compensate for the $\partial_\mu\chi(x)$ terms which will arise when (6.95) is performed on $\mathcal{L}$ of (6.39), we need a vector field $A_\mu(x)$ such that when ϕ changes by (6.95), A_μ changes by

$$A_\mu \to A'_\mu = A_\mu + \partial_\mu\chi. \tag{6.96}$$

In fact, the $\mathcal{L}$ of (6.39) has simply to be replaced by

$$\mathcal{L} = (D_\mu\phi)^\dagger(D^\mu\phi) - V(\phi) - \tfrac{1}{4}F_{\mu\nu}F^{\mu\nu}, \tag{6.97}$$

where D_μ is the $U(1)$ covariant derivative

$$D_\mu = \partial_\mu + \mathrm{i}eA_\mu, \tag{6.98}$$

and as usual $F_{\mu\nu} = \partial_\mu A_\nu - \partial_\nu A_\mu$.

We note at once that we have in this model *four* degrees of freedom – two from A^μ, and two from ϕ_1 and ϕ_2. We anticipate that just as in our toy $A_\mu - \phi$ model of Sections 5.3 and 5.4 (where ϕ was *real*), one of the ϕ's will 'go with the A^μ' to make a massive vector particle, and the other ϕ will turn out to be massive.

We take the same V as (6.40), and argue that for $\mu^2 > 0$ the vacuum will be, *dropping now the suffix* B,

$$\langle 0|\phi|0\rangle = \frac{1}{\sqrt{2}}f, \tag{6.99}$$

where we are thereby selecting $\omega = 0$ in (6.44). Such a field, which has a non-vanishing vacuum expectation value, is now almost universally called a 'Higgs field', in the context of a hidden local symmetry. This $U(1)$ model was first discussed by Higgs (1964*a*, 1964*b*); similar considerations were advanced by Englert & Brout (1964), who treated the non-Abelian case - see the following section - and by Guralnik et al. (1964).

We could proceed as before by setting $\phi_1 = f + \chi_1$, $\phi_2 = \chi_2$ and working $\mathcal{L}$ out. We will do this eventually, but first we note a particularly simple way of looking at the situation. The local gauge transformation on $\phi(x)$ multiplies $\phi(x)$ by an x-dependent phase factor. This suggests that if, instead of decomposing ϕ into two fields ϕ_1 and ϕ_2 via $\phi = (1/\sqrt{2})(\phi_1 - i\phi_2)$, we separated out the 'phase' of ϕ and the 'modulus' of ϕ via

$$\phi(x) = \frac{1}{\sqrt{2}}(f + \rho(x))e^{-i\theta(x)/f}, \tag{6.100}$$

we should be able to get rid of $\theta(x)$ altogether by a suitable choice of $\chi(x)$ in the gauge transformation (6.95). This is indeed so. Under (6.95), we have

$$\phi'(x) = e^{-ie\chi(x)}\phi(x) \tag{6.101}$$

or

$$(f + \rho'(x))e^{-i\theta'(x)/f} = e^{-ie\chi(x)}(f + \rho(x))e^{-i\theta(x)/f}, \tag{6.102}$$

whence

$$\rho'(x) = \rho(x), \theta'(x) = \theta(x) + ef\chi(x). \tag{6.103}$$

So if we choose the particular gauge function:

$$\chi(x) = -\frac{1}{ef}\theta(x), \tag{6.104}$$

we get $\theta' = 0$. Now we recall from Sections 6.3 and 6.6 that the Goldstone modes were precisely the quanta associated with space-time oscillations in the parameter distinguishing the different vacua - which here is the phase angle $\alpha(x) = e\chi(x)$. So they are the quanta of the phase field $\theta(x)$. But in the gauge (6.104) we have $\theta' = 0$, and so such massless excitations will not now be visible in the physical particle spectrum. In the global $U(1)$ case, α was a constant and the Goldstone field could not be cancelled away at all space-time points.

In this gauge, A_μ becomes

$$A'_\mu(x) = A_\mu(x) - \frac{1}{ef}\partial_\mu\theta(x) \tag{6.105}$$

(cf. equation (5.24)!), and of course

$$\phi'(x) = \frac{1}{\sqrt{2}}(f + \rho(x)). \tag{6.106}$$

$\mathcal{L}$ of (6.97) then becomes

$$\mathcal{L} = (\tfrac{1}{2}(\partial_\mu\rho)(\partial^\mu\rho) - \tfrac{1}{2}\mu^2\rho^2) + (-\tfrac{1}{4}F_{A'\mu\nu}F_{A'}{}^{\mu\nu} + \tfrac{1}{2}e^2f^2A'_\mu A'^\mu)$$
$$+ A'_\mu A'^\mu \tfrac{1}{2}e^2(\rho^2 + 2f\rho) - \tfrac{1}{8}\lambda^2\rho^4 - \tfrac{1}{2}\lambda^2 f\rho^3 + \text{constant} \tag{6.107}$$

and $\theta(x)$ has vanished. The local phase fluctuation does not give rise to observable massless scalar particles; instead, it has been absorbed into the electromagnetic field by a gauge transformation, and this field has acquired a longitudinal component $(\partial_\mu\theta)$, becoming massive (cf. the term $\tfrac{1}{2}e^2f^2A'_\mu A'^\mu$ in (6.107). We note from (6.99) that f has the dimension of mass). The four degrees of freedom with which we started have become three for the massive spin-1 A'^μ, and one for the (massive) scalar field ρ. The quanta of fields such as ρ are called 'Higgs bosons'.

This form of $\mathcal{L}$ exhibits the *particle spectrum* very nicely, but it is not convenient for discussing renormalisation, because of all the troubles with the high energy behaviour of the A'-propagator. But our experience with the toy model of Section 5.3 suggests that the way to deal with this is to exploit precisely the *gauge-invariance* of the theory. We anticipate that in a different gauge the high energy behaviour would be better (though the particle interpretation will be less clear). So, we look at an arbitrary gauge, setting

$$\phi_1 = f + \chi_1, \phi_2 = \chi_2\,. \tag{6.108}$$

We find

$$\mathcal{L} = (\tfrac{1}{2}(\partial_\mu\chi_1)(\partial^\mu\chi_1) - \tfrac{1}{2}\mu^2\chi_1{}^2)$$
$$+ (\tfrac{1}{2}(\partial_\mu\chi_2)(\partial^\mu\chi_2) - efA^\mu\partial_\mu\chi_2 - \tfrac{1}{4}F_{\mu\nu}F^{\mu\nu} + \tfrac{1}{2}e^2f^2A_\mu A^\mu)$$
$$+ eA^\mu(\chi_2\partial_\mu\chi_1 - \chi_1\partial_\mu\chi_2) + \tfrac{1}{2}e^2A_\mu A^\mu(\chi_1{}^2 + \chi_2{}^2 + 2f\chi_1)$$
$$- \frac{|\mu|\lambda}{2}\chi_1(\chi_1{}^2 + \chi_2{}^2) - \tfrac{1}{8}\lambda^2(\chi_1{}^2 + \chi_2{}^2)^2\,. \tag{6.109}$$

We see that the *second term in parentheses in* (6.109) *is precisely our toy* (5.26), *with a mass $M = ef$, but now it is nested into a non-trivial interacting field theory.* We know that, thanks to the $A^\mu\partial_\mu\chi_2$ coupling, we can't straightforwardly read off the A-propagator; indeed, under an infinitesimal gauge transformation

$$A_\mu \to A_\mu + \partial_\mu\epsilon(x), \tag{6.110}$$

we find

$$\chi_1(x) \to \chi_1(x) - e\epsilon(x)\chi_2(x) \tag{6.111}$$
$$\chi_2(x) \to \chi_2(x) + e\epsilon(x)f + e\epsilon(x)\chi_1(x), \tag{6.112}$$

so that (6.112) exhibits the *inhomogeneous* term just as in (5.25). Both A_μ and χ_2 experience an inhomogeneous translation. We need to *fix the gauge* in order to quantise, and as before we add the 't Hooft term

$$-\frac{1}{2\xi}\,(\partial_\mu A^\mu + \xi M\chi_2)^2\,,\ \text{where}\ M = ef. \tag{6.113}$$

This then leads to the ξ-dependent propagators

$$\left[-g_{\mu\nu} + \frac{(1-\xi)k_\mu k_\nu}{k^2 - \xi M^2}\right]/(k^2 - M^2),\ \text{for the}\ A\ \text{field} \tag{6.114}$$

and

$$(k^2 - \xi M^2)^{-1},\ \text{for the}\ \chi_2\ \text{field.} \tag{6.115}$$

The χ_1 field is an ordinary scalar field of mass μ. The ξ-dependent poles of (6.114) and (6.115) must always cancel.

Looking at (6.114), we see that, as expected, the A-propagator has improved high energy behaviour (as long as $\xi \neq \infty$). We therefore *conjecture* that this theory is renormalisable in any gauge with finite ξ. The proof is supplied by 't Hooft (1971*b*). The $\xi = \infty$ gauge corresponds to the Lagrangian (6.107) - the χ_2 field disappears, and the A-particle is a conventional massive spin-1 particle with mass M and propagator as in (5.6). This gauge, in which the particle spectrum is easily interpretable, is often called the unitary gauge, or U-gauge. There are no ξ-dependent poles in this unitary gauge, but the theory is not manifestly renormalisable. Gauges such that ξ is finite are called renormalisable gauges, or R-gauges. We rely on the gauge-invariance to let us have our cake (unitarity) and eat it (renormalisability) i.e. one gauge exhibits one desirable property, and another the other, but by gauge-invariance it is all the same theory. Naturally this is not a proof.

The final $\mathcal{L}(\chi_1, \chi_2)$ of (6.109) has no trace of the symmetry remaining, *except* that there are remarkably few independent parameters: only μ, e and f are used to describe the A field, the χ_1 field, and the four interactions $A^2\chi^2, A_\mu\chi\partial_\mu\chi, \chi^3$ and χ^4. This 'conspiracy among the coupling constants' is in fact just what is required to bring about all the necessary cancellations of the bad high energy behaviour of the Born graphs, and (presumably) guarantee renormalisability.

At the end of Section 6.6 we saw how the current of a spontaneously broken global symmetry coupled the vacuum to the Goldstone boson. What happens in the Higgs case? The electromagnetic current corresponding to the *global* symmetry (6.94) of the Lagrangian (6.97), which includes the gauge field, is (as already discussed in Section 3.1, (3.17))

$$j^\mu = ie(\phi^\dagger \partial^\mu \phi - (\partial^\mu \phi)^\dagger \phi) - 2e^2 A^\mu \phi^\dagger \phi. \tag{6.116}$$

Applying (6.108) and retaining only linear terms,

$$j^\mu = -e^2 f^2 A^\mu + ef\partial^\mu \chi_2 + \text{non-linear terms.} \tag{6.117}$$

Using $\partial_\mu j^\mu = 0$ to eliminate χ_2, we find (Taylor 1978)

$$j^\mu = -M^2 \left(g^{\mu\nu} - \frac{\partial^\mu \partial^\nu}{\Box} \right) A_\nu + \text{terms of order } e \tag{6.118}$$

whence

$$\langle 0 | j^\mu | \epsilon, k \rangle = -M^2 \left(g^{\mu\nu} - \frac{k^\mu k^\nu}{M^2} \right) \epsilon_\nu + \text{terms of order } e \tag{6.119}$$

for an on-mass-shell vector particle of mass M and polarisation ϵ. So the current in this case connects the vacuum to the massive vector particle.

An interesting physical interpretation of the origin of the mass M can be obtained by considering the form of the current (6.116). The crucial consequence of the hypothesis of a non-zero vacuum expectation value for ϕ is that there is a non-vanishing current j^μ, even in the vacuum. Further, this vacuum current is proportional to the electromagnetic potential A^μ. If we choose the gauge (6.99) with $\langle 0|\phi|0\rangle$ real, we have

$$\langle 0 | j^\mu | 0 \rangle = -e^2 f^2 A'^\mu = -M^2 A'^\mu, \tag{6.120}$$

where, in accordance with the notation of (6.105), we have denoted the potential by A'^μ, which satisfies the gauge condition $\partial \cdot A' = 0$. The Maxwell equation for the field A'^μ *in vacuo* then reads

$$\Box A'^\mu = -M^2 A'^\mu \tag{6.121}$$

and the components of A'^μ have mass M.

In non-relativistic physics (for static fields) the relation

$$\mathbf{j} = -M^2 \mathbf{A} \tag{6.122}$$

occurs in the theory of superconductivity, where it is known as the London equation (Tilley & Tilley 1974). In that case, the current (6.122) is interpreted as a 'screening current', associated with the electron (Cooper) pairs, whose condensate forms the superconducting ground state. For static fields, (6.122) implies

$$\nabla^2 \mathbf{A} = M^2 \mathbf{A} \tag{6.123}$$

and hence, for one dimensional geometry, the field dies off over a distance of order M^{-1}, called the penetration length. An external magnetic field will only penetrate into the body of a superconductor by a distance of order M^{-1}: this phenomenon is called the Meissner effect. It is a remarkable transference of these ideas – well established in the superconducting case – to suppose that similar 'screening currents' are also present in the particle physics vacuum. Yet this seems to be the only way in which renormalisable massive vector theories can be made (see the following section). Certainly, 'nothing' is turning out to be rather complicated.

The non-relativistic analogy is developed further in Aitchison & Hey (1982). But we add one further comment about it here. The role of the Higgs field is played by the Cooper pair wavefunction, which acts as a classical field, as it describes a condensate (cf. Bogoliubov's treatment of the superfluid ground state). In other words, it is some kind of 'bound state' wavefunction. This raises the possibility that the same might be true of a Higgs field ϕ - which, so far, we have introduced in a quite phenomenological way. According to this view of the matter, one would start from a Lagrangian with no explicit Higgs fields, and hope to demonstrate that some kind of 'pairing' took place, leading to a hidden symmetry phase. So far, no realistic calculations of this sort exist.

A rather different idea is that the symmetry breakdown, for example, in our simple $U(1)$ model, the transition between negative and positive μ^2 in (6.40)), may not be present at the classical level, but is induced by radiative (quantum) corrections. Coleman & Weinberg (1973) first discussed a model of this type - just (6.97) with $\mu^2 = 0$. In this case both the scalar (Higgs) and gauge particles can acquire mass, but their ratio is now *fixed*, in terms of the charge parameter e. Clearly, the model has one fewer parameter than (6.97) and is therefore more predictive.

It goes without saying, of course, that the entire content of this section has been *purely* model building, of no direct relevance to particle physics since, presumably, the real photon is massless. We now take the final step in this long chapter, and introduce a hidden gauge symmetry which *is* of physical relevance.

6.10 Hidden local non-Abelian symmetry

Instead of attempting a general treatment, we shall confine ourselves to the non-Abelian analogue of what we have just done for the specific $U(1)$ Goldstone model. That is, we introduce gauge fields so as to make the global $SU(2) \times U(1)$ group of (6.56) into a local one. We use the covariant derivative to get an $\mathcal{L}$ which is locally invariant:

$$\mathcal{L} = (D_\mu\phi)^\dagger(D^\mu\phi) + \tfrac{1}{2}\mu^2\phi^\dagger\phi - \tfrac{1}{2}\lambda^2(\phi^\dagger\phi)^2 - \tfrac{1}{4}\mathbf{F}_{\mu\nu}\cdot\mathbf{F}^{\mu\nu} - \tfrac{1}{4}G_{\mu\nu}G^{\mu\nu}, \tag{6.124}$$

where the isotriplet of gauge fields $W_\mu{}^a (a = 1, 2, 3)$ are associated with the local $SU(2)$, with $\mathbf{F}^{\mu\nu}$ as in (3.45), and where

$$G_{\mu\nu} = \partial_\mu B_\nu - \partial_\nu B_\mu \tag{6.125}$$

involves the $U(1)$ gauge field $B_\mu(x)$. In (6.124), D_μ is defined by

$$D_\mu\phi = \left(\partial_\mu + \frac{ig}{2}\,\boldsymbol{\tau}\cdot\mathbf{W}_\mu + \frac{ig'}{2}\,B_\mu\right)\phi, \tag{6.126}$$

the straightforward generalisation of (3.34) (the factor $\frac{1}{2}$ in the g' term is put in for later convenience).

As before, we look for the ground state of this $\mathcal{L}$. We cannot allow $\langle 0|\mathbf{W}_\mu|0\rangle \neq 0$, or $\langle 0|B_\mu|0\rangle \neq 0$ or else Lorentz invariance fails; hence only $\langle 0|\phi|0\rangle \neq 0$ can be examined. As before, we set

$$\langle 0|\phi|0\rangle = \begin{pmatrix} 0 \\ f/\sqrt{2} \end{pmatrix} \tag{6.127}$$

but this time we can choose the direction of the isospin axes *independently* at each $\boldsymbol{x}$, so as to align an arbitrary ϕ along the 'down' direction, by doing *different* $SU(2)$ rotations at each space-time point. That is to say, the local $SU(2)$ rotations allow us to view an arbitrary isospinor as a 'down' spinor with respect to suitable axes which, however, differ at each space-time point, as illustrated in Fig. 6.7 (cf. Fig. 6.6). So an arbitrary ϕ can be regarded as a pure phase away from a 'down' isospinor. Thus we may set

$$\phi(x) = \exp(-\mathrm{i}\boldsymbol{\theta}(x)\cdot\boldsymbol{\tau}/2f)\cdot\begin{pmatrix} 0 \\ f/\sqrt{2}+\sigma(x)/\sqrt{2} \end{pmatrix} \tag{6.128}$$

(cf. (6.100) for the $1/f$ in the exponent) and expect to be able to gauge away the $\boldsymbol{\theta}$ field. Indeed, under a local $SU(2)$ transformation,

$$\phi \to \phi' = \exp(-\mathrm{i}g\boldsymbol{\alpha}(x)\cdot\boldsymbol{\tau}/2)\phi,$$

so

$$\boldsymbol{\theta}(x) \to \boldsymbol{\theta}(x) + fg\boldsymbol{\alpha}(x) \tag{6.129}$$

and if we *choose* $\boldsymbol{\alpha} = -\boldsymbol{\theta}/fg$ we get $\boldsymbol{\theta}'(x) = 0$, and the field is then just (dropping the prime)

$$\phi = \begin{pmatrix} 0 \\ (f+\sigma(x))/\sqrt{2} \end{pmatrix}. \tag{6.130}$$

Fig. 6.7. Heavy arrows represent the direction and magnitude of ϕ in isospin space. The dotted lines represent the 'old' internal space axes used in Fig. 6.6. The 'new' internal axes can be chosen independently at each x, t point.

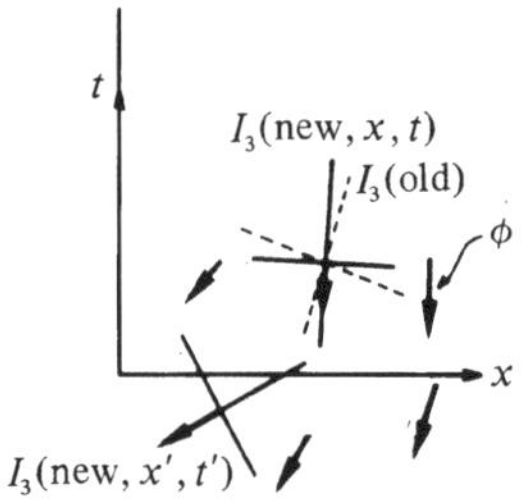

Substituting (6.130) into $\mathcal{L}$ and retaining second-order terms as usual, we find

$$\mathcal{L} = \tfrac{1}{2}\partial_\mu\sigma\partial^\mu\sigma - \tfrac{1}{2}\mu^2\sigma^2 + \tfrac{1}{8}g^2f^2(W_\mu{}^1W^{1\mu} + W_\mu{}^2W^{2\mu})$$
$$+ \tfrac{1}{8}f^2(gW_\mu{}^3 - g'B_\mu)(gW^{3\mu} - g'B^\mu)$$
$$+ \text{higher-order terms} + \text{kinetic terms for } \mathbf{W}, B \text{ fields.} \tag{6.131}$$

We see that although $W_\mu{}^1$ and $W_\mu{}^2$ appear straightforwardly, the fields $W_\mu{}^3$ and B_μ are mixed. But they are easily unmixed by introducing†

$$Z_\mu = \cos\theta_W W_\mu{}^3 - \sin\theta_W B_\mu \tag{6.132}$$
$$A_\mu = \sin\theta_W W_\mu{}^3 + \cos\theta_W B_\mu$$

with

$$\tan\theta_W = g'/g. \tag{6.133}$$

We then find that the 'non-interaction' terms in $\mathcal{L}$ are, finally,

$$\mathcal{L}_{\text{free}} = \tfrac{1}{2}\partial_\mu\sigma\partial^\mu\sigma - \tfrac{1}{2}\mu^2\sigma^2$$
$$-\tfrac{1}{4}(\partial_\mu W_\nu{}^1 - \partial_\nu W_\mu{}^1)(\partial^\mu W^{1\nu} - \partial^\nu W^{1\mu}) + \tfrac{1}{8}g^2f^2(W_\mu{}^1 W^{1\mu})$$
$$-\tfrac{1}{4}(\partial_\mu W_\nu{}^2 - \partial_\nu W_\mu{}^2)(\partial^\mu W^{2\nu} - \partial^\nu W^{2\mu}) + \tfrac{1}{8}g^2f^2(W_\mu{}^2 W^{2\mu})$$
$$-\tfrac{1}{4}(\partial_\mu Z_\nu - \partial_\nu Z_\mu)(\partial^\mu Z^\nu - \partial^\nu Z^\mu) + \tfrac{1}{8}f^2 Z_\mu Z^\mu(g^2 + g'^2)$$
$$-\tfrac{1}{4}(\partial_\mu A_\nu - \partial_\nu A_\mu)(\partial^\mu A^\nu - \partial^\nu A^\mu). \tag{6.134}$$

In this form the physical interpretation is clear: there are three massive vector particles, of masses

$$M_1 = M_2 = \tfrac{1}{2}fg \equiv M_W \tag{6.135}$$
$$M_3 = \tfrac{1}{2}f(g^2 + g'^2)^{1/2} = M_1/\cos\theta_W \equiv M_Z \tag{6.136}$$

and one massless vector particle, as well as one massive scalar particle (the 'Higgs meson' σ, of mass μ). Let us count degrees of freedom. Originally in (6.124) there were three massless W's and one massless B, which is 8 d.f.'s, plus four scalar fields, which is a total of 12 d.f.'s. In the form (6.134), in which the symmetry is hidden, there are three massive vector particles, which makes 9 d.f.'s, and one massless vector particle, making 11 d.f.'s, and one massive scalar making 12 in all again. We emphasise that the masslessness of A_μ (required for its eventual interpretation as the electromagnetic vector potential) follows from the choice of the breaking condition $\langle 0|\phi|0\rangle \neq 0$, which was preserved by the generator combination $\tfrac{1}{2}(1 + \tau_3)/2$, and thus left one vector particle massless (by analogy with the considerations of Section 6.7; see Taylor (1978), Section 6.5).

This interpretation of A_μ is indeed plausible from the form of $D_\mu\phi$ written in terms of A_μ and Z (disregarding the other bits):

† The notation anticipates the application to the GSW theory in the following chapter (c.f. (7.15) and (7.16)), where θ_W is the *weak mixing angle.*

$$D_\mu\phi = \partial_\mu + ig\sin\theta_W \frac{(1+\tau_3)}{2} A_\mu + \frac{ig}{\cos\theta_W} \times \left(\frac{\tau_3}{2} - \sin^2\theta_W \frac{(1+\tau_3)}{2}\right) Z_\mu + \ldots \tag{6.137}$$

In (6.137) A_μ appears correctly coupled to the conventional *charge* operator $(1+\tau_3)/2$, which is the preserved symmetry generator. We also note that if this interpretation is correct physically, $g\sin\theta_W$ will be identified with the physical charge e. This brings us, at last, to the application of these ideas to the electroweak interactions, which we shall discuss in the following chapter.

Finally, after the preparation of Section 6.9 it should now be straightforward to understand that the form (6.134) corresponds to a *particular choice of gauge* (that given by $\boldsymbol{\alpha} = -\boldsymbol{\theta}/fg$, cf. (6.129)) - it is the unitary gauge in the terminology of Section 6.9. There is always the possibility of choosing other gauges, as in the Abelian case, and this will, in general, be advantageous for *renormalisation* questions. We would then return to the general parametrisation

$$\phi = \begin{pmatrix} 0 \\ f/\sqrt{2} \end{pmatrix} + \frac{1}{\sqrt{2}} \begin{pmatrix} \phi_2' + i\phi_1' \\ \sigma - i\phi_3' \end{pmatrix} \tag{6.138}$$

and add 't Hooft gauge-fixing terms

$$\frac{-1}{2\xi}\left\{ \sum_{\alpha=1,2} (\partial_\mu W^{\alpha\mu} + \xi M_W \phi'^{\alpha})^2 + (\partial_\mu Z^\mu + \xi M_Z \phi_3') + (\partial_\mu A^\mu)^2 \right\}, \tag{6.139}$$

leading to ξ-dependent propagators as in (6.114) and (6.115). In such gauges, the Feynman rules will of course have to include graphs corresponding to exchange of quanta of the unphysical fields ϕ' as well as those of the physical Higgs field σ-

The general non-Abelian case has been treated by Kibble (1967).

7 Theory of weak and electromagnetic interactions

7.1 Introduction

In 1960, Glashow, pursuing the goal† of unifying the electromagnetic and weak interactions, proposed a 'partially symmetric' unified (electroweak) theory. By a 'partial symmetry' he meant one which was broken only by explicit mass terms in the Lagrangian (as in PCAC, discussed in Section 6.8). Considering the then-known charged weak currents, and of course the electromagnetic current, the natural candidate for a unified theory would be one involving three vector bosons, as in Section 5.2 above - the $W^{\pm}$, and the γ. It was clearly realised, though, that M_W could not be zero, a fact which, to quote Glashow (1961), constituted 'the principal stumbling block in any pursuit of the analogy between hypothetical vector mesons and photons'. Nevertheless, he went on: 'It is a stumbling block we must overlook.' One must marvel at such intuition, not to say faith. He went on to show that the simplest partially symmetric theory of the weak and electromagnetic interactions, in which the symmetry was violated only by explicit M_W terms in the Lagrangian, had to involve a *neutral* weak current, in addition to the charged ones then known. This implied the introduction of four bosons in all, the $W^{\pm}$, the γ, and the neutral one Z^0. In this scheme the symmetry group was precisely the $SU(2) \times U(1)$ of Sections 6.7 and 6.10. As a weak interaction symmetry, however, it applied to *leptons* (as well as to hadrons), and accordingly the leptons were to be classified according to representations of this 'weak' isospin and 'weak' hypercharge. Because M_W (and M_Z) were so large, no visible residue of such a symmetry was likely to remain in the observed lepton spectrum.

This proposal still left unanswered the question of how the mass terms were to be treated, and also, in particular, the question of renormalisability. It was Weinberg (1967) and, independently, Salam (1968), who suggested that making the gauge symmetry hidden could provide the resolution of these difficulties. They proposed that the vector bosons in the Glashow model could acquire masses via the mechanism detailed in Section 6.10. On this view it is essential

† Early contributions in this quest were made by Schwinger (1957) and Bludman (1958).

that the $SU(2) \times U(1)$ symmetry be a *local* one, the vector particles being the quanta of the gauge fields, so that, as it was hoped, enough of the gauge-invariance might be 'retained' in the broken phase to render the theory renormalisable. In such a situation, the symmetry will manifest itself, as we have repeatedly stressed, in relations between the parameters of the theory (masses and coupling constants) rather than in any pattern of approximate mass multiplets. These relations will be such, it may be hoped, as to cause just those cancellations among the Born graphs of the theory as are required to produce high energy behaviour consistent with renormalisability (cf. Section 5.2). Renormalisability was proved by 't Hooft (1971*b*).

7.2 The GSW model (leptons only)

We proceed, then, with the development of the GSW theory, as applied to leptons only, for the moment. The Lagrangian of Section 6.10 included only the gauge fields and the (Higgs) scalar field ϕ – it was, of course, deliberately set up to be the appropriate Lagrangian for that sector of the more complete physical situation we are now going to discuss. In particular we shall see that A_μ *is* to be identified with the real electromagnetic field, and the $W^\pm$ and Z *are* to be the quanta associated with the charged and neutral weak currents respectively. How, then, do we include leptons? Let us denote by (t, t_3) the quantum numbers of the weak isospin group $SU(2)$, and by y the weak 'hypercharge' associated with the $U(1)$ group. Clearly, the W's have $t=1$, and we must employ

$$W = \frac{1}{\sqrt{2}}(W^1 - \mathrm{i}W^2), \quad W^\dagger = \frac{1}{\sqrt{2}}(W^1 + \mathrm{i}W^2), \tag{7.1}$$

associated with $t_3 = \pm 1$. W will *destroy* a W^+, or *create* a W^-. (We have already assigned $t(\phi) = \frac{1}{2}$; for $y(\phi)$, see below.) A basic process involving a lepton pair is, for example,

$$W^- \to e^- + \bar{\nu}_e \tag{7.2}$$

so that, given that $t(W) = 1$, a simple possibility is to put e^- and ν_e into a weak isodoublet

$$\ell_e = \begin{pmatrix} \nu_e \\ e^- \end{pmatrix}. \tag{7.3}$$

Then we expect that a *globally* $SU(2)$-invariant interaction will be $\bar{\ell}_e \tau/2 \ell_e \cdot \mathbf{W}$ (we do the Lorentz structure later). We can check that if

$$\mathbf{j} = \bar{\ell}_e \boldsymbol{\tau}/2 \ell_e, \tag{7.4}$$

then

$$\mathbf{j} \cdot \mathbf{W} = (\bar{\nu}_e e^- W + \bar{e} \nu_e W^\dagger) + j_3 W^3, \tag{7.5}$$

in which, for example, the term $\bar{e}\nu_e W^\dagger$ destroys a W^- and creates an e^- and a $\bar{\nu}_e$. We do the same thing for the μ-like leptons

$$\ell_\mu = \begin{pmatrix} \nu_\mu \\ \mu^- \end{pmatrix} \tag{7.6}$$

and for any other lepton 'generations' that may be needed.

Considering now the Lorentz structure, we know - see for example, Marshak et al. (1969) - that the weak leptonic currents have pure $V-A$ form, and so we are actually dealing with currents of the type

$$\bar{e}\gamma_\mu(1-\gamma_5)\nu_e W^{\dagger\mu} \tag{7.7}$$

ignoring the isospace labels now. Since under the internal $SU(2)$ group we are contemplating arbitrary unitary transformation of e^-'s and ν_e's, while the neutrino field, it appears from experiment, has only the left-handed (L) helicity component, we cannot allow the electron field in (7.3) to contain both helicities. We therefore take the basic isodoublet to be, *not* (7.3), but

$$\ell_e = \begin{pmatrix} \dfrac{(1-\gamma_5)}{2}\nu_e \\ \dfrac{(1-\gamma_5)}{2}e^- \end{pmatrix}, \tag{7.8}$$

where, of course, since (we believe) $\gamma_5\nu_e = -\nu_e$, it follows that $\frac{1}{2}(1-\gamma_5)\nu_e = \nu_e$. (7.8) is then the weak left-handed isodoublet of electron-type leptons, and it can be written alternatively as

$$\ell_e = \begin{pmatrix} \nu_e \\ e_L{}^- \end{pmatrix}, \quad e_L{}^- = \frac{(1-\gamma_5)}{2}e^-. \tag{7.9}$$

Our global $SU(2)$ transformations on (7.8) are now

$$\delta\begin{pmatrix} \nu_e \\ e_L{}^- \end{pmatrix} = -\mathrm{i}\boldsymbol{\epsilon}\cdot\boldsymbol{\tau}/2\,\frac{(1-\gamma_5)}{2}\begin{pmatrix} \nu_e \\ e^- \end{pmatrix}, \tag{7.10}$$

i.e. it is an '$SU(2)_L$' (recall Section 6.8) in which the projection operators $\frac{1}{2}(1-\gamma_5)$ are included. (7.10) of course implies

$$\delta\ell_e = -\mathrm{i}\boldsymbol{\epsilon}\cdot\boldsymbol{\tau}/2\,\ell_e \tag{7.11}$$

and associated global currents

$$\mathbf{j}_\mu = \bar{\ell}_e\gamma_\mu\frac{\boldsymbol{\tau}}{2}\ell_e,$$

with ℓ_e given by (7.8). If we consider a similar transformation on the 'right-handed' doublet

$$r_e = \begin{pmatrix} \dfrac{(1+\gamma_5)}{2}\nu_e \\ \dfrac{(1+\gamma_5)}{2}e^- \end{pmatrix}, \tag{7.12}$$

we find $\delta(r_e) = 0$. Actually, if $\gamma_5 \nu_e = -\nu_e$, the top member of (7.12) is zero identically - but in any case r_e is a *singlet* under the global transformation. We shall often omit the subscript 'L' from $SU(2)_L$, where it is easily understood.

Now consider the $U(1)$ part. To be consistent with (6.126), we take the $SU(2)_L \times U(1)$ covariant derivative to be

$$\partial_\mu + igt \cdot W_\mu + \tfrac{1}{2}ig'yB_\mu, \tag{7.13}$$

in which (suppressing all unnecessary symbols) the t_3, y part is

$$\partial_\mu + igt_3 W_\mu{}^3 + \tfrac{1}{2}ig'yB_\mu. \tag{7.14}$$

But from (6.132) and (6.133)

$$W_\mu{}^3 = \sin\theta_W A_\mu + \cos\theta_W Z_\mu \tag{7.15}$$

$$B_\mu = \cos\theta_W A_\mu - \sin\theta_W Z_\mu, \tag{7.16}$$

so that the A_μ part of (7.13) is

$$\partial_\mu + i(gt_3 \sin\theta_W + \tfrac{1}{2}g'y\cos\theta_W)A_\mu. \tag{7.17}$$

But since from (6.133) $g'\cos\theta_W = g\sin\theta_W$, (7.17) becomes

$$\partial_\mu + ig\sin\theta_W(t_3 + \tfrac{1}{2}y)A_\mu, \tag{7.18}$$

allowing us to interpret '$t_3 + \frac{1}{2}y$' as the *charge operator*, with A_μ the electromagnetic gauge field, whence (with e defined to be *positive*)

$$g\sin\theta_W = e. \tag{7.19}$$

We arrive at the quantum number assignments shown in Table 7.1 - note that (6.126) has taken $y(\phi) = 1$ (or else g' in (6.136) would have to be changed).

With this assignment we can now make the theory *locally* $SU(2) \times U(1)$-invariant, by simply writing down the covariant derivatives for all the relevant

Table 7.1. $Q = t_3 + \frac{1}{2}y$

	t	t_3	y	Q
ν_e, ν_μ	$\frac{1}{2}$	$\frac{1}{2}$	-1	0
e_L, μ_L	$\frac{1}{2}$	$-\frac{1}{2}$	-1	-1
e_R, μ_R	0	0	-2	-1
Higgs field	$\frac{1}{2}$	$\pm\frac{1}{2}$	1	1, 0

representations (including a possibly redundant ν_R):

$$D_\mu \ell_e = \left(\partial_\mu + ig\frac{\boldsymbol{\tau}\cdot\mathbf{W}_\mu}{2} - \tfrac{1}{2}ig' B_\mu\right)\ell_e \qquad (7.20)$$

$$D_\mu e_R = (\partial_\mu - ig' B_\mu)e_R \qquad (7.21)$$

$$D_\mu \nu_R = \partial_\mu \nu_R, \qquad (7.22)$$

where $e_R = [\frac{1}{2}(1+\gamma_5)]e^-$. The weak interaction of leptons and gauge fields is now *determined* by postulating that the leptons have *only* those interactions induced by the replacement $\partial_\mu \to D_\mu$ in the free Lagrangian. However, there is one snag before we can achieve a fully gauge-invariant Lagrangian. It is easy to see that any lepton mass term is *not* invariant under $\psi \to [\frac{1}{2}(1-\gamma_5)]\psi$ (cf. Section 6.8). To make this important point quite clear, let us decompose ψ as

$$\psi = \frac{(1+\gamma_5)}{2}\psi + \frac{(1-\gamma_5)}{2}\psi \equiv \psi_R + \psi_L; \qquad (7.23)$$

then we find that a fermion mass term can be written as

$$m\bar{\psi}\psi = m(\bar{\psi}_R\psi_L + \bar{\psi}_L\psi_R), \qquad (7.24)$$

since $\bar{\psi}_R\psi_R = \bar{\psi}_L\psi_L = 0$, using $\gamma_5\gamma_0 = -\gamma_0\gamma_5$, and $(1-\gamma_5)(1+\gamma_5) = 0$. Since ψ_L and ψ_R transform differently under $SU(2)_L$, (7.24) is not, in fact, globally $SU(2)_L$-invariant. Hence the inclusion of lepton mass terms will also spoil *local* $SU(2)_L$-gauge-invariance, and therefore prejudice renormalisability. We postpone the treatment of this problem for the moment, since we are keen to arrive at some experimental consequences after so much theory – these consequences will be relevant to a regime where the lepton masses are negligible.

Looking then at the lepton and gauge field part of the electroweak Lagrangian constructed according to the gauge principle $\partial_\mu \to D_\mu$, we find the terms (for e's and μ's only)

$$
\begin{aligned}
& i\bar{\ell}_e \not{D}\ell_e + i\bar{\ell}_\mu \not{D}\ell_\mu + i\bar{e}_R \not{D} e_R + i\bar{\mu}_R \not{D}\mu_R + i\bar{\nu}_{eR}\not{\partial}\nu_{eR} + i\bar{\nu}_{\mu R}\not{\partial}\nu_{\mu R} \\
&\quad = \text{free part} \\
&\quad -g/\sqrt{2}\left(\bar{\nu}_e\gamma^\mu \frac{(1-\gamma_5)}{2} e^- W_\mu + \text{Herm. conj.}\right) \\
&\quad + e\,\bar{e}\gamma^\mu e A_\mu \\
&\quad - \frac{e}{\sin\theta_W \cos\theta_W}\left(\tfrac{1}{2}\bar{\nu}_e\gamma^\mu\nu_e - \tfrac{1}{2}\cos 2\theta_W \bar{e}_L\gamma^\mu e_L\right. \\
&\quad \left. + \sin^2\theta_W \bar{e}_R\gamma^\mu e_R\right)Z_\mu + \text{muon parts.}
\end{aligned} \qquad (7.25)
$$

The $g/\sqrt{2}$ term is the conventional IVB charged current term, the e term is the ordinary electromagnetic current term (with $Q = -1$ for the e^-, of course), and the last term is the *neutral weak current* term $(-g/\cos\theta_W)j^{Z\mu}Z_\mu$, in this model.

7.3 Some experimental consequences

(1) Connection with phenomenological four-fermion theory: vector boson masses

If we use the charged current of (7.25) to calculate say μ^- decay, to leading order, we obtain an amplitude which, neglecting the momentum dependence of the W-propagator is

$$\frac{1}{M_W{}^2}\left(\frac{g}{\sqrt{2}}\right)^2 \tfrac{1}{2}\bar{u}_{\nu_\mu}\gamma^\mu(1-\gamma_5)u_{\mu^-}\tfrac{1}{2}\bar{u}_{e^-}\gamma_\mu(1-\gamma_5)v_{\nu_e}. \tag{7.26}$$

This is an interaction of the original Fermi (1934*a*, 1934*b*) type, and we may identify the Fermi coupling constant G_F as

$$\frac{G_F}{\sqrt{2}} = \frac{g^2}{8M_W{}^2}. \tag{7.27}$$

Using (7.19), we find

$$M_W = \frac{e}{2^{5/4}G_F{}^{1/2}\sin\theta_W} \approx \frac{37}{\sin\theta_W}\ \text{GeV}. \tag{7.28}$$

As we shall see below, the experimentally determined value of $\sin\theta_W$ seems to be slightly less than $\frac{1}{2}$; the associated prediction† for M_W is then ~78 GeV. From (6.136), the theory predicts† $M_Z = M_W/\cos\theta_W \sim 89$ GeV.

(2) Width of the W

The width of the W to lepton pairs can easily be calculated to leading order, and comes out to be

$$\Gamma_{W\to e\nu} = 23\ \text{MeV}/\sin^3\theta_W. \tag{7.29}$$

Including all modes, and setting $\sin\theta_W \sim \frac{1}{2}$, might give a total W width of a few GeV. The experimental observation of a W 'resonance' of this sort of width, and the predicted mass, will surely be a spectacular triumph for the gauge principle approach.

(3) ν–e scattering

There are three reactions of interest,

$$\nu_\mu e^- \to \nu_\mu e^- \tag{7.30}$$

$$\bar{\nu}_\mu e^- \to \bar{\nu}_\mu e^- \tag{7.31}$$

$$\nu_e e^- \to \nu_e e^-. \tag{7.32}$$

† These values (and the value of θ_W extracted from experiments using lowest-order formulae) will be changed by higher-order electroweak corrections (Sirlin 1980, Marciano & Sirlin 1981, Llewellyn Smith & Wheater 1981). The discovery of the $W^\pm$ is reported in Arnison et al. (1983*a*) and Banner et al. (1983), and of the Z^0 in Arnison et al. (1983*b*) and Bagnaia et al. (1983).

The first two proceed (in lowest-order – 'tree' – approximation) by Z^0 exchange only Fig. (7.1) while the last can also go by W^- exchange (Fig. 7.2). The first two processes are therefore characteristically *weak neutral current effects*, in lowest order. The effective interaction (at energies large compared to m_e but small compared to M_Z) is, from (7.25),

$$-g^2 \frac{1}{M_Z^2 \cos^2 \theta_W} j_\mu{}^Z j^{Z\mu}, \tag{7.33}$$

the minus sign coming from the i's in the vertices and propagator, where $j_\mu{}^Z$ is the term in brackets in (7.25) – the $(\cos\theta_W)^{-1}$ factors being taken out in (7.33). The ν contribution to $j_\mu{}^Z$ gives

$$\tfrac{1}{2}\bar{\nu}\gamma_\mu\nu. \tag{7.34}$$

The electron contribution is

$$\{-\tfrac{1}{2}\cos 2\theta_W \bar{e}_L\gamma_\mu e_L + \sin^2\theta_W \bar{e}_R\gamma_\mu e_R\}, \tag{7.35}$$

which can be re-written as

$$\tfrac{1}{2}\{\bar{e}\gamma_\mu e C_V + \bar{e}\gamma_\mu\gamma_5 e C_A\}, \tag{7.36}$$

with

$$C_V = -\tfrac{1}{2} + 2\sin^2\theta_W \tag{7.37}$$

$$C_A = +\tfrac{1}{2}, \tag{7.38}$$

or, of course, as

$$\{\bar{e}_L\gamma_\mu e_L C_L + \bar{e}_R\gamma_\mu e_R C_R\}, \tag{7.39}$$

with

$$C_L = -\tfrac{1}{2} + \sin^2\theta_W, \quad C_R = +\sin^2\theta_W. \tag{7.40}$$

Fig. 7.1. Z^0 exchange graph in $\nu_\mu(\bar{\nu}_\mu)e^- \rightarrow \nu_\mu(\bar{\nu}_\mu)e^-$.

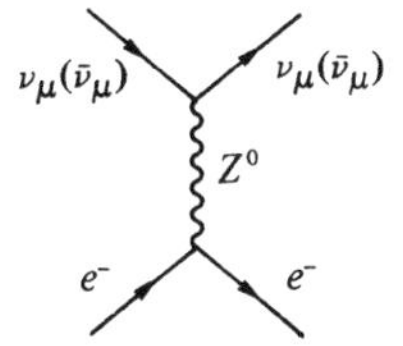

Fig. 7.2. Z^0 and W^- exchange graphs in $\nu_e e^- \rightarrow \nu_e e^-$.

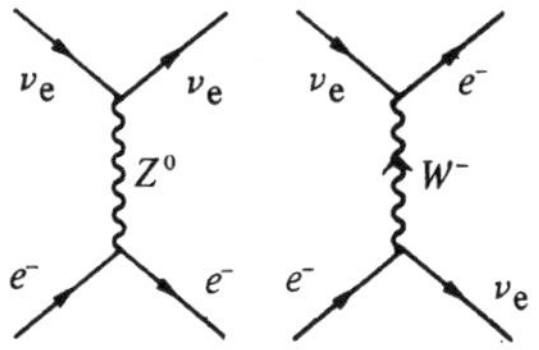

(7.33) is therefore

$$-2\sqrt{2}G_F\bar{\nu}\gamma_\mu\nu\{\bar{e}_L\gamma^\mu e_L C_L + \bar{e}_R\gamma^\mu e_R C_R\}. \qquad (7.41)$$

The cross-section for (7.30) was calculated (for non-zero electron mass) by 't Hooft (1971*c*) to be

$$\frac{d\sigma}{dy} = \frac{2G_F{}^2 m_e E_\nu}{\pi} \times \{|C_L|^2 + |C_R|^2(1-y)^2 - \tfrac{1}{2}(C_R{}^*C_L + C_L{}^*C_R)ym_e/E_\nu\}, \qquad (7.42)$$

where $y = E_e'/E_\nu$, and E_e' is the electron recoil energy in the laboratory system. We have quoted (7.42) with the ym_e/E_ν term included: this is negligible for $E_\nu \gg m_e$ and is not derivable in our massless theory of course, (see Section 7.4 below for the inclusion of masses). The cross-section for the antineutrino process (7.31) is obtained by reversing the helicity labels, $C_L \leftrightarrow C_R$. Integrating over the y values (from 0 to 1) we find for neutrinos

$$\sigma_\nu = \frac{2G_F{}^2 m_e E_\nu}{\pi}\{|C_L|^2 + \tfrac{1}{3}|C_R|^2\}, \qquad (7.43)$$

where

$$\frac{G_F{}^2 m_e}{\pi} \sim 0.8 \times 10^{-41}\ \text{cm}^2/\text{GeV}; \qquad (7.44)$$

and for antineutrinos

$$\sigma_{\bar{\nu}} = \frac{2G_F{}^2 m_e E_{\bar{\nu}}}{\pi}\{|C_R|^2 + \tfrac{1}{3}|C_L|^2\}, \qquad (7.45)$$

with C_L and C_R given by (7.40). Similar calculations can be done for the process (7.32).

The experimental situation is frequently reviewed. We will briefly mention some other, not purely leptonic, processes below, but the above model accounts excellently for the present data (Kim et al. 1981), with a value† of the parameter θ_W in the region

$$\sin^2\theta_W \approx 0.225 \qquad (7.46)$$

with an error of about ±0.015.

(4) Parity violation effects

One characteristic *class* of predictions of the $SU(2)_L \times U(1)$ theory is that there will be parity violation effects in 'apparently' electromagnetic processes, due to interference between γ- and Z^0-exchange graphs. For example,

† But note the footnote on p. 110, and also the discussion following (7.81) below.

the colliding beam reaction

$$e^+e^- \to \mu^+\mu^- \tag{7.47}$$

can go via an intermediate state of one photon, but also a state of one Z^0. The effect will be proportional to a $C_V C_A$ interference term, and hence (from (7.37)) necessarily somewhat suppressed due to the empirical near-equality $2\sin^2\theta_W \approx \frac{1}{2}$. However, the relative contribution of the Z^0 rises with centre of mass energy E, and for $E \geqslant 20$ GeV the work of Budny (1973), for example, shows that parity violation effects from γ-Z interference could be detectable. These effects include muon (or, perhaps more easily, tau) longitudinal polarisation, and muon (or tau) asymmetry from polarised electrons. Other parity violation effects which have been measured involve processes which are not purely leptonic, and will be mentioned in Section 5 below, along with further experimental consequences of the theory.

7.4 A more complete Lagrangian

The terms (7.25) were enough to enable us to discuss the electroweak interactions of the leptons and gauge fields, but of course the full Lagrangian contains many other terms. We may list the terms we know about already:

(i) the free Yang-Mills Lagrangian for the W's (cf. 3.49)), $-\frac{1}{4}\mathbf{F}_{\mu\nu}\cdot\mathbf{F}^{\mu\nu} \equiv \mathcal{L}_W$;

(ii) and for the B, $-\frac{1}{4}F_{B\mu\nu}F_B{}^{\mu\nu} \equiv \mathcal{L}_B$;

(iii) the lepton part $(7.25) \equiv \mathcal{L}_{\text{leptons}}$;

(iv) the Higgs sector part (cf. (6.124), (6.126)) $(D_\mu\phi)^\dagger(D^\mu\phi) + \frac{1}{2}\mu^2\phi^\dagger\phi - \frac{1}{2}\lambda^2(\phi^\dagger\phi)^2 \equiv \mathcal{L}_\phi$;

(v) and in general a gauge-fixing part as in (6.139), call it $\mathcal{L}_G$.

We may now consider how to give the leptons mass. It must be done in such a way as not to destroy the global and local gauge-invariance. The basic idea is to introduce a Yukawa-type coupling of the form

$$g_s\bar{\psi}\psi\phi \tag{7.48}$$

between a lepton field ψ and the Higgs field ϕ. Then, when the shift (6.130) (or, in the more general gauge, (6.138)) is done, we will collect a term of the form

$$g_s\bar{\psi}\psi f/\surd 2 \tag{7.49}$$

as well as interaction terms; (7.49) has the form of a lepton mass term.

Of course, this is precisely the mechanism adopted in the σ-model of Section 6.8. We recall from (7.24) that an actual lepton mass term will have the form $m(\bar{\psi}_R\psi_L + \bar{\psi}_L\psi_R)$, so that ϕ has to couple to these particular combinations in an $SU(2)_L \times U(1)$-invariant manner. Referring to Table 7.1, we see that $\bar{e}_R$ is a weak isosinglet and creates $y = -2$, ℓ_e is a weak isodoublet destroying $y = -1$, and $\phi^\dagger$ is an isodoublet creating $y=+1$. The combination

$$g_s(\bar{e}_R\phi^\dagger \ell_e + \text{Herm. conj.}) \tag{7.50}$$

is therefore globally (and locally) $SU(2)_L \times U(1)$-invariant. If we now choose $g_s = -\sqrt{2}m_e/f$, (7.50) will produce, after the shift (6.130), the required electron mass term

$$-m_e(\bar{e}_R e_L + \bar{e}_L e_R), \tag{7.51}$$

to which the analogous μ^- terms (and heavier lepton terms) can be added similarly.

The parameter f in the foregoing (we recall that $f/\sqrt{2}$ is the vacuum expectation value $|\mu|/(\sqrt{2}|\lambda|)$ of ϕ) is related to G_F by (cf. (6.135) and (7.27))

$$f = \frac{1}{2^{1/4}G_F{}^{1/2}} \approx 246\,\text{GeV}. \tag{7.52}$$

It is not, we emphasise, predicted by the foregoing theory. The coupling strength g_s is therefore

$$|g_s| = \sqrt{2}m_e f^{-1} \approx 3 \times 10^{-6} \tag{7.53}$$

for electrons, and some 200 times larger for muons; and correspondingly for more massive leptons. The theory as so far developed does not *predict* these lepton masses either, of course, nor the mass μ of the ϕ field. Indeed, non-zero masses for the neutrinos could be incorporated gauge-invariantly if desired. All these masses, and also the quark masses which will appear when the theory is extended to include hadrons (see Section 5), must be accepted as free parameters at present (though larger group structures are capable of providing relations between some of them - see Section 6).

7.5 The inclusion of hadrons

Hadrons are introduced by including *quark* field terms in the Lagrangian; these terms will actually be rather similar to the lepton terms. How the quarks then 'make' hadrons - presumably via their QCD interactions - we will not discuss here. The hope is that a simple electroweak theory might exist for the 'elementary' quarks, though presumably not for the 'composite' hadrons. At all events, the quarks must be introduced into the Lagrangian in a way which preserves the local gauge-invariance (and thus renormalisability). This is a rather powerful constraint. It means that the part of the Lagrangian involving quarks must have exhibited a global symmetry under the $SU(2)_L \times U(1)$ group which was *exact* before it was gauged (made local), and before the broken phase was induced by $\langle 0|\phi|0\rangle \neq 0$. The Noether currents of this global symmetry are, as we have seen for leptons (see after (7.11), and compare with (7.25)), the quantities which couple to the gauge fields, and have the interpretation of the *weak currents for quarks.* This allows derivation of selection rules based simply on the quantum numbers of these currents, and also (using the parton model in the deep inelastic regime) a direct test of electroweak quark couplings.

However, there is a difficulty with a too naive introduction of flavour-carrying quarks. The most obvious way of incorporating them leads to *neutral, strangeness-changing weak currents* of strength comparable to the other experimentally well-established currents. But there are very strict upper limits on the strengths of such neutral strangeness-changing currents, as indicated by the decay branching ratios, and hence there is a difficulty. The argument is as follows. To accommodate the quark fields in the lepton model discussed so far, we must introduce quark analogues of the weak leptonic currents $\mathbf{j}^{\mu}$ and $j^{y\mu}$, which we shall call $\mathbf{J}^{\mu}$ and $J^{y\mu}$. We already know quite a lot about the quark currents from phenomenology. The ordinary charged current is believed to have the structure (see, for example, Bailin 1977)

$$\bar{d}'\gamma_{\mu}\tfrac{1}{2}(1-\gamma_5)u + \text{Herm. conj.}, \tag{7.54}$$

where in a world of only three flavours (so far) the quark field is

$$q = \begin{pmatrix} u \\ d \\ s \end{pmatrix}, \tag{7.55}$$

in which u, d are the members of a (*hadronic*!) isospin doublet, and s is an isospin singlet, and u, d have strangeness zero, while s has strangeness -1. Finally, d' in (7.54) is defined by

$$d' = d\cos\theta_{\mathrm{C}} + s\sin\theta_{\mathrm{C}} \tag{7.56}$$

in terms of the Cabibbo (1963) angle θ_{C} ($\sin\theta_{\mathrm{C}} \approx 0.22$). It will be useful also to introduce the orthogonal combination

$$s' = -d\sin\theta_{\mathrm{C}} + s\cos\theta_{\mathrm{C}}. \tag{7.57}$$

(7.54) can be unpacked into terms like

$$\cos\theta_{\mathrm{C}}\bar{u}\gamma_{\mu}\tfrac{1}{2}(1-\gamma_5)d + \sin\theta_{\mathrm{C}}\bar{u}\gamma_{\mu}\tfrac{1}{2}(1-\gamma_5)s, \tag{7.58}$$

in which the first piece acts in $n \to pe^-\bar{\nu}_e$ and the second in $\Sigma^- \to ne^-\bar{\nu}_e$. Now (7.54) can be regarded as part of an $SU(3)_f$ (f for flavour) *octet* of currents which, following Taylor (1978), we shall denote by $L_{\mu}^{\prime\alpha}$

$$L_{\mu}^{\prime\alpha} = \bar{q}'\gamma_{\mu}\frac{\lambda^{\alpha}}{2}\frac{(1-\gamma_5)}{2}q', \tag{7.59}$$

where

$$q' = \begin{pmatrix} u \\ d' \\ s' \end{pmatrix} \tag{7.60}$$

and the λ^{α} are the usual matrices of $SU(3)_f$.

In (7.59) the notation L is used because of the left-handed helicity projection operator $\frac{1}{2}(1-\gamma_5)$. We can also introduce the corresponding right-handed currents

$$R'^{\alpha}_{\mu} = \bar{q}'\gamma_\mu \frac{\lambda^\alpha}{2}\frac{(1+\gamma_5)}{2}q'. \tag{7.61}$$

The associated charges, call them $Q_L{}^\alpha$ and $Q_R{}^\alpha$, generate the algebra of $SU(3)_{fL} \times SU(3)_{fR}$ - the generalisation of the $SU(2)_L \times SU(2)_R$ of Section 6.8. For further details, see Adler & Dashen (1968).

The question now is, how to associate the phenomenological currents L'^{α}_{μ} with the quark currents of the weak $SU(2)_L \times U(1)$ scheme of the previous section. The obvious move would be to *identify* the $SU(2)_{fL}$ subgroup part of (7.59) as the required $SU(2)_L$ quark currents:

$$?\, J_\mu{}^i \equiv L'^{i}_{\mu} \quad (i = 1, 2, 3), \tag{7.62}$$

a choice which would certainly couple the charged W's correctly to the empirically correct quark current L'^{i}_{μ}. It would also seem reasonable to identify the $U(1)$ part as

$$\tfrac{1}{2}J_\mu{}^y = J_\mu{}^{\mathrm{em}} - L'^{3}_{\mu}, \tag{7.63}$$

where $J_\mu{}^{\mathrm{em}}$ is the electromagnetic current of quarks. However, it is (7.63) that leads to the difficulty: L'^{3}_{μ} can be expanded as

$$L'^{3}_{\mu} = (1 - \tfrac{1}{2}\sin^2\theta_C)L_\mu{}^3 + \frac{\sqrt{3}}{2}\sin^2\theta_C L_\mu{}^8 - \cos\theta_C \sin\theta_C L_\mu{}^6, \tag{7.64}$$

in which $L_\mu{}^6$ is a neutral operator which includes a strangeness-changing component. In terms of the quark fields,

$$L_\mu{}^6 = \frac{1}{2}\left[\bar{d}\gamma_\mu \frac{1}{2}(1-\gamma_5)\, s + \bar{s}\gamma_\mu \frac{1}{2}(1-\gamma_5)\, d\right] \tag{7.65}$$

Hence the Z^0, coupling to $J_\mu{}^y$, would in this model mediate neutral strangeness-changing weak decays of an unacceptable strength.

A way of overcoming this difficulty, while preserving the gauge framework, was suggested by Glashow, Iliopoulos & Maiani (1970) - though they were also guided, at least in part, by a desire to see some *lepton-quark symmetry* (see further below). GIM introduced a fourth quark c carrying a new flavour, called charm $C(c) = +1$, and having charge 2/3 e. In this scheme, one has then four leptons

$$\{\nu_e, e^-, \nu_\mu, \mu^-\}$$

with charges (0, −1, 0, −1), and four quarks u, d, c, s with the 'shifted' charges $(\frac{2}{3}, -\frac{1}{3}, \frac{2}{3}, -\frac{1}{3})$. In the lepton model, we dealt with *two* leptonic isodoublets

$$\begin{pmatrix} \nu_e \\ e_L^- \end{pmatrix}, \quad \begin{pmatrix} \nu_\mu \\ \mu_L^- \end{pmatrix}.$$

On the basis of the charge analogy emphasised by GIM, we could imagine associating ν_e with u, and ν_μ with c. Having done this, however, we could then associate e^- with d and μ^- with s, *or* e^- with s and μ^- with d. Indeed, a general choice of the form

$$\left.\begin{aligned} e^- &\to d' = \cos\theta_C d + \sin\theta_C s \\ \mu^- &\to s' = -\sin\theta_C d + \cos\theta_C s \end{aligned}\right\} \tag{7.66}$$

could, on grounds of this 'symmetry' alone, be envisaged. In any case, the GIM proposal was that one should consider two weak quark isodoublets

$$q_u = \begin{pmatrix} u \\ d' \end{pmatrix}_L, \quad q_c = \begin{pmatrix} c \\ s' \end{pmatrix}_L \tag{7.67}$$

as the fundamental quark structures relevant for the construction of the $SU(2)_L \times U(1)$ currents, by analogy with the leptons. The corresponding right-handed quark fields would all be $SU(2)_L$ singlets, as usual. In this case, if we preserve the relation

$$Q = t_3 + \tfrac{1}{2}y \tag{7.68}$$

(cf. (7.18)), we find $y(u) = \frac{1}{3} = y(d') = y(c) = y(s')$. The $SU(2)_L \times U(1)$ quark currents are then

$$\mathbf{J}_\mu = \bar{q}_u \frac{\boldsymbol{\tau}}{2} \gamma_\mu q_u + \bar{q}_c \frac{\boldsymbol{\tau}}{2} \gamma_\mu q_c \tag{7.69}$$

and

$$J_\mu{}^y = \bar{q}_u \gamma_\mu q_u + \bar{q}_c \gamma_\mu q_c \cdot + \text{contribution from RH SU(2)}_L \text{ singlets} \tag{7.70}$$

The neutral current $J_\mu{}^3$ is

$$J_\mu{}^3 = \tfrac{1}{2}(\bar{u}_L \gamma_\mu u_L - \bar{d}_L \gamma_\mu d_L + \bar{c}_L \gamma_\mu c_L - \bar{s}_L \gamma_\mu s_L) \tag{7.71}$$

and does not change strangeness. The Z^0 is coupled to a linear combination of this current and the electromagnetic current (which does not change strangeness either) and so there will be no neutral $\Delta S \neq 0$ processes *to lowest order* in this theory.

We can now proceed as in the lepton model to write down a globally, and then locally, $SU(2)_L \times U(1)$-invariant theory involving the quark fields in the combinations (7.67). It will exactly parallel the lepton case, so the results can be read off. The charged W's will couple via

$$-\frac{g}{\sqrt{2}} \bar{u} \gamma_\mu \frac{(1 - \gamma_5)}{2} d' W^\mu + \text{Herm. conj.} \tag{7.72}$$

(essentially by construction) but we shall also have *new* couplings of the form

$$-\frac{g}{\sqrt{2}}\bar{c}\gamma_\mu\frac{(1-\gamma_5)}{2}s'W^\mu + \text{Herm. conj.}, \tag{7.73}$$

which were predicted by GIM. Note that the *same* g appears in both the quark and lepton interactions - an illustration of the 'universality' of gauge interactions (cf. Section 3.3, comment 5 and Section 8.2 below).

It is not our purpose here to provide a review of weak interactions, but we shall for completeness briefly mention the salient experimental consequences of the GIM theory. First and foremost, of course, is the *prediction* of a new hadronic quantum number, *charm*, whose existence has now been amply confirmed (Goldhaber et al. 1976; Peruzzi et al. 1976; Feldman et al. 1977). The identification of charmed particles is made particularly clear by the definite *selection rules in charm-changing decays* which (7.73) predicts. Consider the *semi-leptonic* decay of a charmed hadron into non-charmed hadrons (the decay being thought of entirely in quark field terms). Since Z^0 processes are diagonal in charm, only W-mediated processes contribute. The current is as in (7.73), which will induce decays

$$c \to s + \ell^+ + \nu_e \qquad \text{with strength } \cos\theta_C$$

$$c \to d + \ell^+ + \nu_e \qquad \text{with strength } -\sin\theta_C.$$

Since $\sin\theta_C \ll \cos\theta_C$, we find $\Delta Q = \Delta C \approx \Delta S$ for these decays. So if a charmed particle is produced semi-leptonically from an initially $C=0$ state (for example, in a ν interaction), and then decays semi-leptonically to a $C=0$ state, we may apply $\Delta Q = \Delta C$ to each of the production and decay stages to predict that the total dilepton charge will be zero (for the data, see Benvenuti et al. 1975*a*, 1975*b*, 1975*c*; Barish et al. 1976). Now consider the *non-leptonic* decay, for example of a charmed meson D^+ having the composition $(c\bar{d})$ $(Q=+1, C=+1, S=0)$. This will couple, with strength $\cos\theta_C$, to $(s\bar{d}) + W^+$ via (7.73), and the W^+ will couple to $C=0$ hadrons via (7.72), to $S=0$ states with strength $\cos\theta_C$ and to $S\neq 0$ states with strength $\sin\theta_C$. Hence the *dominant* mode is expected to be $\Delta S = \Delta C$, with amplitude $\cos^2\theta_C$, $\Delta S = 0$ having amplitude $\cos\theta_C \sin\theta_C$ and $\Delta S = -\Delta C$ amplitude $\sin^2\theta_C$. So in D^+ decay we expect

$$D^+ \to K^-\pi^+\pi^+ \tag{7.74}$$

$$C=1 \quad S=-1$$

but *not*

$$D^+ \to K^+\pi^-\pi^+ \tag{7.75}$$

$$C=1 \quad S=+1.$$

This is just what is observed (Peruzzi et al. 1976).

The masses of charmed mesons are of order $\geqslant 1.8$ GeV, indicating that the mass of the charmed quark is of order 1.5 GeV (taking a simple-minded view of constituent masses). This is within the range *predicted* by GIM from the following interesting argument. By analogy with the lepton couplings (7.13), and using (7.15) and (7.16) with $g' \cos\theta_W = g \sin\theta_W$, we obtain the quark current coupling to Z^0 as

$$J_\mu{}^Z = \cos^2\theta_W J_\mu{}^3 - \tfrac{1}{2}\sin^2\theta_W J_\mu{}^y \tag{7.76}$$

(which has strength $g/\cos\theta_W$, as in (7.25)), which can also be re-written using (7.68) as, for example

$$J_\mu{}^Z = J_\mu{}^3 - \sin^2\theta_W J_\mu{}^{\text{em}}. \tag{7.77}$$

Now we have seen that $J_\mu{}^3$ is strangeness-non-changing, and so is $J_\mu{}^{\text{em}}$, so that to *first order* in G_F there are no neutral strangeness-changing processes. But the observed fraction

$$\frac{\Gamma(K_L \to \mu^+\mu^-)}{\Gamma(K_L \to \text{all})} = (9.1 \pm 1.8)\ 10^{-9} \tag{7.78}$$

is so low as to put a limit on a *second-order* contribution. $K_L \to \mu^+\mu^-$ can proceed via the processes shown in Figs. 7.3*a* and 7.3*b*. The particular way in which θ_C enters into the couplings (7.72) and (7.73) ensures that if $m_u = m_c$ *these two graphs will exactly cancel.* Calculation shows that the amplitude is of order

$$\sin\theta_C \frac{g^4}{M_W{}^2}(m_c{}^2 - m_u{}^2)/M_W{}^2, \tag{7.79}$$

so that, in order not to arrive at too high an estimate for $\Gamma(K_L \to \mu^+\mu^-)$, m_c cannot be arbitrarily larger than m_u. GIM estimated $m_c \sim 1$ to 3 GeV. The remarkable properties of the J/ψ particles (Aubert et al. 1974; Augustin et al. 1974) are now reasonably-well understood by regarding them as $c\bar{c}$ composites, with $m_c \sim 1.5$ GeV, the value predicted by Gaillard and Lee (1974).

It is, of course, fundamental to the $SU(2)_L \times U(1)$ theory as extended to quarks that there must be observable weak neutral-current phenomena, in weak

Fig. 7.3. Two-W intermediate state graphs contributing to $\Delta S \neq 0$ weak neutral transition $K_L{}^0 \to \mu^+\mu^-$: (a) u exchange graph; (b) c exchange graph.

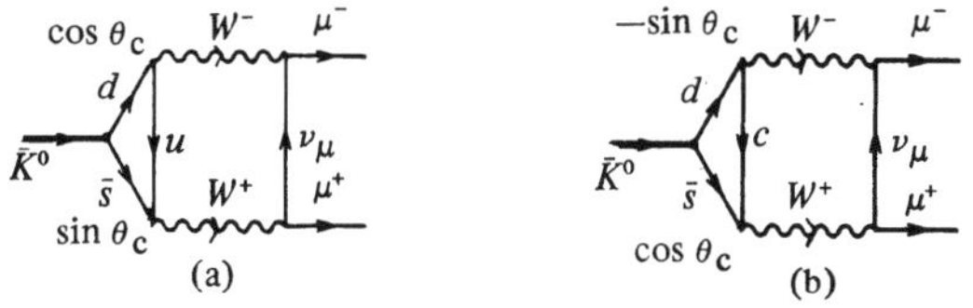

interactions of hadrons, mediated by the Z-current (7.76). Indeed, such phenomena were first seen in inelastic neutrino scattering from nucleons (Hasert et al. 1973; Benvenuti et al. 1974; Barish et al. 1975) and in single pion production by neutrinos from nucleons (Barish et al. 1974; Lee et al. 1977). Such experiments all require for their interpretation in terms of quark currents some additional assumptions - such as, for example, those of the parton model (Close 1979) for inelastic scattering. An especially interesting class of predictions concerns parity violation effects in apparently electromagnetic interactions of hadrons, due to Z^0-γ interference (the hadronic analogue of the effect discussed in Section 3 above). Parity violation effects in atomic physics, due to this process, were first discussed by Bouchiat & Bouchiat (1974). The experimental situation is becoming more settled, and supports the theory at least qualitatively. Reviews are provided by Fortson & Wilets (1980), and by Cummins & Bucksbaum (1980).

At present, the clearest evidence on this sort of effect is provided by the classic experiment on the inelastic scattering of longitudinally polarised electrons from deuterium at SLAC (Prescott et al. 1978). In this experiment the flux of inelastically scattered electrons was measured for incident electrons of definite helicity. An asymmetry between the results for the two different helicities was observed, a clear signal for parity violation. The asymmetry A has been calculated for various gauge models (see, for example, Cahn & Gilman 1978), and is of the order of $10^{-4}Q^2$, where Q^2 is the invariant momentum transfer, measured in GeV^2. (The quark-paton model, or some alternative, also has to be assumed in these calculations.) The simplest $SU(2)_L \times U(1)$ model is in good agreement with the measured A/Q^2, with an angle $\sin^2 \theta_W = 0.20 \pm 0.03$, consistent with the values obtained from the neutrino experiments. A limited amount of data was also taken with a hydrogen target, with consistent results, although with a larger error. Neutral-current experiments have been reviewed and analysed by Kim et al. (1981).

Once again, quark masses have to be introduced by hand via an $SU(2)_L \times U(1)$-invariant Yukawa coupling of quarks to the Higgs field, which generates the masses after induced symmetry breakdown. No relations between the quark masses are required in such a model.

7.6 More flavours; generations; 'Grand Unification'

We first note that in the theory as so far formulated, with four flavours, there is no obviously 'natural' way of accommodating CP violation (though attempts have been made to introduce it via a hidden symmetry mechanism, as reviewed by Mohapatra 1978). However, a six-flavour model can accommodate CP violation quite simply (Maskawa & Kobayashi 1973), although it cannot be said to predict it.

But there are, of course, other more compelling reasons to think that quarks come in more than four flavours. The properties of the upsilon (Υ) family appear likely to repeat those of the ψ's at a higher energy, strongly suggesting an interpretation of the states as $b\bar{b}$ bound states, where b is a quark of a fifth flavour. The lepton-quark symmetry is preserved by the discovery of the τ lepton (Perl et al. 1976). It is expected that the latter must have its own neutrino ν_τ, forming a weak isodoublet with τ_L. The b quark then needs a t quark to complete the pattern, and we have a six-flavour theory.

The different columns in the resulting scheme

$$\begin{array}{cccc} \begin{pmatrix} u \\ d' \end{pmatrix}_L & \begin{pmatrix} c \\ s' \end{pmatrix}_L & \begin{pmatrix} t \\ b' \end{pmatrix}_L & \ldots ? \\ \begin{pmatrix} \nu_e \\ e^- \end{pmatrix}_L & \begin{pmatrix} \nu_\mu \\ \mu^- \end{pmatrix}_L & \begin{pmatrix} \nu_\tau \\ \tau^- \end{pmatrix}_L & \ldots ? \end{array} \tag{7.80}$$

are called generations. No compelling reason for such duplication is yet known.

It may be salutary at this point to count up the number of parameters in this type of electroweak theory of quarks and leptons. *Assuming* that (for no good reason) the neutrinos are all exactly massless, there are nine masses, four generalised Cabibbo angles (describing the mixing of the quark fields, one of them corresponding to a CP-violating phase), two gauge coupling constants g, g', and two parameters for the Higgs potential (at least). There may also be θ parameters† (Belavin et al. 1975; 't Hooft 1976, 1978), of the type mentioned in Section 3.4.

It is fair to say that a very great deal of progress has been made in even arriving at this stage; and the character of the questions that are now being raised is of a qualitatively different type from earlier, more phenomenological, preoccupations. Nevertheless, a theory with 17 or more parameters cannot be regarded as the last word. For this reason alone it seems promising to try and embed the $SU(2)_L \times U(1)$ group into a larger group, in particular one which does not have an Abelian factor (cf. Section 3.3, comment 7) - i.e. into a *simple* group.

A further and very appealing reason for considering a larger symmetry group is the possibility of including within it the colour group $SU(3)_C$ of QCD as well. It is a remarkable vindication of the local symmetry point of view that it does now seem very likely that the weak and strong interactions, like the well-established electromagnetic one, are gauge field theories. This opens the way to a 'Grand' unification of the $SU(3)_C \times SU(2)_L \times U(1)$ structure within a larger simple group. Detailed discussion of various unification schemes is outside the scope of this Introduction. However, we shall briefly mention a few of the characteristic consequences, using the group $SU(5)$ (Georgi & Glashow 1974) as an example.

† Not to be confused with the weak mixing angle.

In any such theory, with a single gauge group, there will be only *one* gauge coupling constant. Consequently, relations will be imposed on the constants associated with the $SU(3)_C$, $SU(2)_L$ and $U(1)_y$ subgroups; essentially, they will all be related to the single gauge coupling constant via 'Clebsch–Gordan' coefficients. This would be at least some reduction in the number of independent parameters. In particular, the $SU(2)_L$ and $U(1)_y$ constants g and g' will be related, so that a *prediction* will be obtained for the weak angle (cf. (6.133)). In $SU(5)$, the value predicted is (Georgi & Glashow 1974)

$$\sin^2 \theta_W = \tfrac{3}{8}. \tag{7.81}$$

At first sight, this value is discouragingly far from the experimental one (7.46). However, there is an aspect to the concept of 'interaction strength' that we have not yet touched on. This is the idea that higher-order corrections in perturbation theory may be regarded as altering ('renormalising') a coupling constant from the value defined, say, via the lowest-order graphs corresponding to the initial Lagrangian. Furthermore, these changes will be energy-dependent. We shall discuss such effects in more detail in the following chapter, where we shall learn that non-Abelian gauge theories differ from all other renormalisable field theories precisely in the way that the 'effective' coupling constant varies with energy - namely, for the former theories it decreases as the energy increases, tending eventually to zero (a property known as 'asymptotic freedom' - Politzer 1973; Gross & Wilczek 1973). By contrast, the effective coupling constant of a $U(1)$ Abelian gauge group (and all other renormalisable theories) increases with energy.

In the case of $SU(2)_L \times U(1)_y$, g is associated with the non-Abelian $SU(2)_L$, and g' with the Abelian $U(1)_y$, so that the effective g will decrease, while g' will increase, as the energy increases. Since $g'/g = \tan \theta_W$, θ_W will increase with energy. Hence it is possible that (7.81) may accurately represent $\sin^2 \theta_W$ at very high energy, while (7.46) is its measured value at presently accessible energies.

The immediate question is then - how high would the energy have to be in order to 'renormalise' (7.81) into the value (7.46)? We shall see in the following chapter that the energy dependence is typically logarithmic and, consequently, we expect the energy scale required to bring about such a large renormalisation to be very high indeed. The first calculation along these lines by Georgi, Quinn & Weinberg (1974) estimated that the symmetry limit (7.81) would be reached at a mass scale of approximately 10^{14} GeV. Later, more refined, estimates have not changed this value qualitatively (Buras et al. 1978; Ross 1978; Goldman & Ross 1979, 1980; Llewellyn Smith et al. 1981; for a review see Llewellyn Smith 1982*b*).

It seems, in fact, to be characteristic of such unified theories that they involve some similar 'superheavy' mass scale. A further argument leading to the same conclusion is the following. All such schemes contemplate putting quarks and leptons into common multiplets - after all, the differences between the strength of the interactions, which distinguished hadrons from leptons according to the $SU(3)_C \times SU(2)_L \times U(1)$ classification, are supposed to disappear in a view which assigns the same intrinsic strength to all the interactions. Symmetry transformations will then exist, which turn leptons into quarks (or vice versa). When the symmetry group is 'gauged', there will be a gauge quantum for each generator of the group. These quanta mediating quark $\leftrightarrow$ lepton transitions will have to be very massive, in order to suppress the rate for baryon decay, occurring via a process such as that shown in Fig. 7.4. A simple estimate for the proton lifetime τ_p is

$$\tau_p \sim M_X{}^4/\alpha_G{}^2 m_p^5, \qquad (7.82)$$

where α_G is the fine structure constant of the unifying group G, and X is the generic name for any gauge boson mediating a baryon decay. From (7.82) we find $M_X \gtrsim 10^{14}$ GeV, in order not to conflict with the present bound ($\gtrsim 10^{30}$ *yrs*) on τ_p, if $\alpha_G \sim 10^{-2}$. Thus we are led to the same sort of mass scale as before.

One can envisage much more quantitative calculations, which are clearly essential to perform, since the proton lifetime is proportional to $M_X{}^4$. These calculations link M_X, $\sin^2\theta_W$ and τ_p; they also lead to changes (of order 5 GeV) in the simple predictions of M_W and M_Z given earlier, due to radiative corrections. A recent review is provided by Llewellyn Smith (1982*b*). Experiments are being planned to measure, or place improved bounds upon, the proton lifetime. The actual observation of the mode $p \to e^+\pi^0$, for example, would surely be an extraordinary indication that these unified theories are on the right track (see, for example, Weinberg 1981).

We are therefore faced with two very different scales of symmetry breakdown - the first, at extremely high energy ($\sim 10^{14}$ GeV), in which the 'M_X' bosons acquire mass, and the second ($\sim 10^2$ GeV) in which the W's and Z^0 do. No natural explanation for these two disparate scales is yet known.

A further immediate consequence of theories which group quarks and leptons together is that mass ratios will be predicted. For example, $SU(5)$ predicts (Chanowitz et al. 1977; Buras et al. 1978; Nanopoulos & Ross 1979)

Fig. 7.4. Process mediating the decay $p \to e^+\pi^0$.

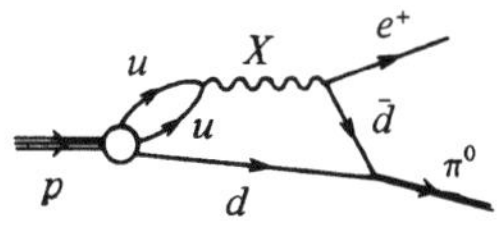

$$m_e = m_d$$
$$m_\mu = m_s$$
$$m_\tau = m_b. \tag{7.83}$$

Here again, the relations (7.83) are only supposed to hold in the symmetry regime of energies $\gg M_X$. Thus large (mass) renormalisation effects may be expected as one comes down to accessible energies. It is claimed (Chanowitz et al. 1977; Buras et al. 1978; Nanopoulos & Ross (1979)) that these effects can be reliably calculated for m_b, with the result $m_b \approx 5$-5.5 GeV (this depends on there being just *six* flavours, the estimate increasing with the number of flavours).

In conclusion, we mention two additional points. The first is that, as remarked earlier, any compact simple group, without a non-compact $U(1)$ factor involving electromagnetism, will lead automatically to charge quantisation. The fractional quark charges are also naturally understood within this general framework, and the particular assignment to representations chosen in the $SU(5)$ scheme (Georgi & Glashow 1974). Secondly, quite independently of any particular gauge theory, it appears that a purely *global* symmetry, such as baryon number conservation seems to be, has no natural place in a thoroughgoing local symmetry framework. The same is true of the lepton numbers (Weinberg 1979; Wilczek & Zee 1979).

8 Renormalisation matters

Renormalisation is so central a topic in quantum field theory - and particularly in gauge field theories - that even an informal introduction to the latter should not omit all mention of it. Indeed, it has already been touched on several times in earlier chapters. In this final chapter, we want to review (while shamelessly borrowing technical details from reliable sources) several important aspects of renormalisation, especially those which impinge particularly on gauge theories.

A renormalisable theory is one in which, by introducing a finite number of parameters taken from experiment, calculations of physical quantities can be carried out, with finite results, up to (in principle) arbitrarily high orders in perturbation theory, by following a well-defined calculational procedure. The main ideas in the procedure will be illustrated for QED in Section 8.1. By contrast, no such procedure has, as yet, been devised for theories which are not renormalisable: in these terms, such theories would require the introduction of an infinite number of arbitrary constants. Of course, we could always fit data with such a theory, but it seems clear that we would have very few real predictions to make. In QED, on the other hand, many predictions can be made, to an accuracy of a few parts per million, given just the values of the fundamental charge e and the electron mass (and possibly masses of other particles); and these predictions have been verified experimentally. At present, then, and until such time as a procedure is invented for obtaining finite physical predictions from theories which are not renormalisable in perturbation theory, the only *calculable* theories are renormalisable ones. This is a good reason for restricting attention to such theories.

In the case of gauge theories, renormalisation-like quantisation itself - involves some particular complications. The symmetries which are fundamental to such theories turn out to imply certain constraints on the amplitudes, namely, identitie of the Ward (1950)-Takahashi (1957) type, and their generalisations. These identities have several roles. They contribute essentially to proving renormalisability in the first place, and they are also necessary to secure a sensible physical interpretation of the theory (they guarantee, for example, that physical amplitudes are independent of the choice of gauge, and that the perturbativelyconstructed Feynman graphs satisfy unitarity - cf. Section 4.4). They also lead to testable predictions (low energy theorems, sum rules) for physical amplitudes. We shall review some of these aspects in Section 8.2.

The renormalisation 'problem' which we have (implicitly) been discussing so far is that of the treatment of the ultraviolet divergences of integrals encountered in perturbation theory. As a first step in analysing these divergences, some procedure has to be adopted which renders the integrals finite, so that sensible manipulations can be performed. Such a procedure is called a 'regularisation'. It amounts to a temporary modification of the theory, which is removed after the manipulations have been done. A very interesting situation can arise if it is not possible to find a regularisation which respects the *symmetries* of the theory. Then, the corresponding Ward identities will, in general, be changed - they will involve 'anomalies'. We shall discuss this in Section 8.3, for a number of different internal symmetries, including the important case of local (gauge) symmetries.

In Section 8.4 we shall treat a different type of symmetry, that of scale invariance, from this point of view. In some ways, scale invariance is rather a detour away from gauge theories. Nevertheless, it is perhaps appropriate to end with a discussion of it, since (as we pointed out in the first chapter) it was precisely from the study of scale invariance that the phase invariance of gauge theories did emerge. Besides, we shall be able to include some mention of the important property of 'asymptotic freedom', possessed uniquely by non-Abelian gauge theories.

8.1 Renormalisation in QED: counter terms

Let us consider a simple Lagrangian of QED type:

$$\mathcal{L} = \bar{\psi}(\mathrm{i}\not{\partial} - m)\psi + e\bar{\psi}\gamma^{\mu}\psi A_{\mu} - \tfrac{1}{4}F_{\mu\nu}F^{\mu\nu} - \frac{1}{2\xi}(\partial\cdot A)^2, \tag{8.1}$$

describing the interaction of a spinor field - call it an electron - of mass m and charge $-e$ interacting with the electromagnetic field. The points we wish to make will be sufficiently illustrated by the calculation of the process shown in Fig. 8.1, the lowest-order correction to the electron propagator. The original propagator itself is

$$\mathrm{i}S_{\mathrm{F}}(p) = \frac{\mathrm{i}}{\not{p} - m} \tag{8.2}$$

Fig. 8.1. Lowest-order virtual photon correction to the electron propagator.

for an electron line carrying four-momentum p^μ, and the correction of Fig. 8.1 adds to (8.2) a term of the form

$$\frac{\mathrm{i}}{\not{p}-m}\,[-\mathrm{i}\Sigma^{[2]}(p)]\,\frac{\mathrm{i}}{\not{p}-m}, \tag{8.3}$$

since it involves initial and final unmodified electron lines. The quantity $-\mathrm{i}\Sigma^{[2]}(p)$ represents the loop part, without these end lines. It is called the electron self energy; the superscript means that we are only calculating the second-order (in powers of e) contribution to it. The expression for it, according to the rules for propagators and vertices following from (8.1), is

$$-\mathrm{i}\Sigma^{[2]}(p) = (+\mathrm{i}e)^2 \int \frac{d^4k}{(2\pi)^4}\frac{-\mathrm{i}}{k^2+\mathrm{i}\epsilon}\gamma_\nu \frac{\mathrm{i}}{\not{p}-\not{k}-m+\mathrm{i}\epsilon}\gamma^\nu \tag{8.4}$$

if we use Feynman gauge $\xi = 1$ in (4.44) - we comment further on this below. The immediate difficulty with (8.4) is that it diverges for large values of k, as is apparent from the fact that there are only three powers of k in the denominator. (We ought also to worry about the fact that it may diverge at the small k end - but this 'infra-red' problem is a quite separate one from renormalisability, and we shall mostly ignore it.)

In order to proceed at all, we have to introduce a regularisation procedure. Probably the most widely-adopted now is the elegant dimensional regularisation method due to 't Hooft & Veltmann (1972*a*); but for the present purposes a simple cut-off is sufficient - for example, we might modify the photon propagator according to

$$\frac{1}{k^2} \to \frac{1}{k^2} - \frac{1}{k^2-\Lambda^2} \tag{8.5}$$

with the large mass Λ eventually being taken to infinity. The resulting Λ-dependent amplitude $\Sigma^{[2]}(p,\Lambda)$ can then be evaluated exactly (Itzykson & Zuber, 1980, Section 7-1-2), with the result that

$$\Sigma^{[2]}(p,\Lambda) = \frac{\alpha}{4\pi}\{3m\ln(\Lambda^2/m^2) - (\not{p}-m)\ln(\Lambda^2/m^2) + \text{part independent of } \Lambda\}. \tag{8.6}$$

We note that $\Sigma^{[2]}$ exhibits a logarithmic divergence, which is a weaker singularity than might have been anticipated from counting powers of k in (8.4). The divergence affects the constant term and the coefficient of the $(\not{p}-m)$ term in (8.6), which can be viewed as an expansion of $\Sigma^{[2]}$ in powers of $(\not{p}-m)$ about the point $\not{p}=m$. The coefficients of all higher powers of $(\not{p}-m)$ are all finite as $\Lambda\to\infty$, since they are given by derivatives of (8.4) with respect to $\not{p}$ and these produce extra powers of k in the denominator, causing convergence.

In addition to the process of Fig. 8.1, we can clearly consider the infinite series indicated in Fig. 8.2, which sums to

$$\frac{\mathrm{i}}{\not{p} - m - \Sigma^{[2]}(p, \Lambda)} \tag{8.7}$$

for the corrected propagator. Graphs such as those in Fig. 8.2 are called 'improper' or ('one-particle reducible'), meaning that they can be divided into two disjoint parts by the removal of one internal fermion line. By contrast, Fig. 8.1 itself is called 'proper' (or 'one-particle irreducible'), since it cannot be so divided. The complete propagator has the general form

$$\mathrm{i}\tilde{S}'_{\mathrm{F}}(p, \Lambda) = \frac{\mathrm{i}}{\not{p} - m - \Sigma(p, \Lambda)} \tag{8.8}$$

where $\Sigma(p, \Lambda)$ is the sum of all proper electron self-energy graphs (i.e. of the form of Fig. 8.1, with the external lines removed). Note that our use of the tilde differs from that in Bjorken & Drell (1965).

Returning to (8.2) and (8.7), we see that whereas the free propagator (8.2) had a pole at $\not{p} = m$, the propagator (8.7) – which is (8.2) as corrected by a certain class of interaction effects – is no longer at $\not{p} = m$. To order α, the pole in (8.7) is at the point

$$\not{p} = m + \delta\tilde{m}^{[2]}(\Lambda) + \text{ part independent of } \Lambda \tag{8.9}$$

where

$$\delta\tilde{m}^{[2]}(\Lambda) = \frac{3\alpha m}{4\pi} \ln(\Lambda^2/m^2). \tag{8.10}$$

Further, whereas the residue of the pole at $p = m$ is i in (8.2), the residue of the pole of (8.7) will not be i. A pole in a propagator represents a point in momentum space at which a particle can be 'on-shell' ($p^2 = m^2$), propagating freely over macroscopic distances. It is natural to associate the result (8.9), then, with a change in the mass of the electron. Such a mass shift is commonplace in solid state physics, where one regularly considers the 'effective mass' of an electron in a solid, modified from its free space value by the effect of the interactions experienced by the electron in the solid. In the present case, the idea is the same, but there are two important differences. The first is that the effect is occurring in free space itself, due to virtual emission and re-absorption effects (quantum fluctuations in the vacuum); the second is that the effect is infinite as $\Lambda \to \infty$. Unless there is some physical cut-off supplied by nature (a fundamental small

Fig. 8.2. Series of sequential corrections of the form of Fig. 8.1.

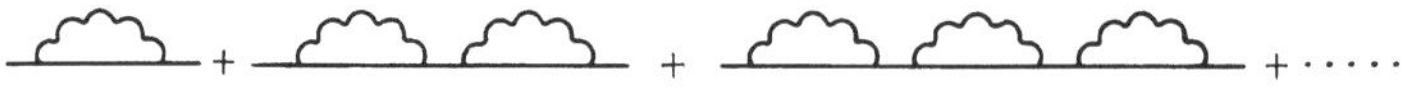

length ? some large mass ?) we must assume that indeed Λ should be allowed to go to infinity, and thus we are forced to do something to make sense of the situation.

Two attitudes may be taken. In the first, we may say that the parameter m in the original Lagrangian (8.1) was not the real mass, as measured in the laboratory - there is no way in which we can remove all these virtual interactions present in the vacuum, and, consequently, the measured mass must be a quantity which includes them all. In the same way, the charge parameter e in (8.1) will become modified into a measured 'effective' charge. Physical amplitudes can then be expressed in terms of these physical parameters instead of ones in the original Lagrangian. The process can be accomplished systematically, order-by-order, in perturbation theory, so that eventually the physical parameters can all be expressed as a series of terms, ordered in powers of the original coupling strength, each of which are formally infinite, although the sum of each series - being a physical parameter - has to be finite. The values of these physical parameters are taken from experiment. In a renormalisable theory, the introduction of a finite number of such parameters renders all other amplitudes finite.

Although this point of view has helpful connections with solid state physics examples, it is not the most suitable for the subsequent development. It is ultimately preferable to consider a somewhat different approach. According to the latter, the parameters appearing in (8.1) are the *physical* ones, but we shall modify (8.1) by the inclusion of *additional* terms, called 'counter terms'. The function of these terms will be to cancel the infinite parts of amplitudes calculated from (8.1) alone, so that physical amplitudes - calculated from (8.1) with the additional counter terms - will all be finite. The counter terms needed for these cancellations have to be calculated, order-by-order (successively), in perturbation theory. We shall discuss the $O(\alpha)$ calculation, and the general question of the definition and interpretation of such terms.

As an example, suppose we added a term

$$\delta\tilde{m}^{[2]}(\Lambda)\bar{\psi}\psi \tag{8.11}$$

to (8.1). Although this has the appearance of a fermion mass term (actually of negative sign, by comparison with the usual Dirac term $-m\bar{\psi}\psi$, see further below), since $\delta\tilde{m}^{[2]}$ depends on α it may be interpreted as an additional interaction; and since we multiply the Lagrangian by i to get the Feynman rule for a vertex, (8.11) contributes a term

$$\frac{\mathrm{i}}{\not{p}-m}\,\mathrm{i}\delta\tilde{m}^{[2]}(\Lambda)\,\frac{\mathrm{i}}{\not{p}-m} \tag{8.12}$$

to the electron propagator, to order α, corresponding to Fig. 8.3. The contributions (8.3) and (8.12) together may then be iterated in the fashion of Fig. 8.2, to yield

$$\frac{\mathrm{i}}{\not{p} - m - \Sigma^{[2]}(p, \Lambda) + \delta\tilde{m}^{[2]}(\Lambda)}. \tag{8.13}$$

The denominator is now

$$\not{p} - m - \left(\delta\tilde{m}^{[2]}(\Lambda) - \frac{\alpha}{4\pi}(\not{p} - m)\ln(\Lambda^2/m^2) + \text{part independent of } \Lambda\right)$$
$$+ \delta\tilde{m}^{[2]}(\Lambda) \tag{8.14}$$

and the *infinite* constant part has been cancelled. In a similar way, the infinite coefficient of the $\not{p} - m$ term in $\Sigma^{[2]}$ may be cancelled by the addition of a term

$$-\frac{\alpha}{4\pi}\ln(\Lambda^2/m^2)\bar{\psi}(\mathrm{i}\not{\partial} - m)\psi \tag{8.15}$$

to (8.1).

We may now note that although, with the addition of the new pieces (8.11) and (8.15), our Lagrangian produces (*for the diagrams considered*) a finite electron propagator, the position of the pole is not in general at $\not{p} = m$, since the 'part independent of Λ' in (8.6) – which is consequently always finite – may be non-zero as $\not{p} \to m$. This in fact is what happens, and it therefore leads to a *finite* shift (renormalisation) in the mass parameter m of (8.1). Now there is nothing sacrosanct about the terms (8.11) and (8.15) except that they must cancel the indicated infinities. We can perfectly well, if we choose, insert some additional finite contribution into (8.11) or (8.15), to suit our convenience. It is clearly rather convenient (though not essential) to arrange matters in these counter terms so that the parameter m in (8.1) actually *is* the position of the pole in the complete electron propagator, i.e. it is the physical electron mass. We may accomplish this by replacing (8.11) by

$$\delta m^{[2]}(\Lambda)\bar{\psi}\psi, \tag{8.16}$$

such that

$$\Sigma^{[2]}(p, \Lambda) - \delta m^{[2]}(\Lambda) = 0 \tag{8.17}$$

for $\not{p} = m$. The calculation of Itzykson & Zuber (1980) gives

$$\delta m^{[2]}(\Lambda) = \delta\tilde{m}^{[2]}(\Lambda) + \frac{3\alpha m}{8\pi}, \tag{8.18}$$

showing clearly the finite change in the counter term.

Fig. 8.3. Contribution of the counter term (8.11) to the electron propagator.

Similarly, we may add a finite part to (8.15), if we wish, thereby altering the *residue* of the pole in the propagator at $\not{p}=m$. We can choose this finite part so that, for example, the residue has the 'free particle' value, namely i. Conventionally, the counter term (8.15) is written

$$(Z_2^{[2]}(\Lambda)-1)\bar{\psi}(\mathrm{i}\not{\partial}-m)\psi, \tag{8.19}$$

so that the propagator to this order is, with counter terms (8.16) and (8.19) included,

$$\frac{\mathrm{i}}{\not{p}-m-\Sigma^{[2]}(p,\Lambda)+\delta m^{[2]}(\Lambda)+(Z_2^{[2]}(\Lambda)-1)(\not{p}-m)}. \tag{8.20}$$

The demand that the residue at $\not{p}=m$ be i fixes $Z_2^{[2]}$:

$$Z_2^{[2]}(\Lambda)-1=\frac{\partial\Sigma^{[2]}}{\partial\not{p}}(p,\Lambda)\bigg|_{\not{p}=m} \tag{8.21}$$

Clearly we will find $Z_2^{[2]}=1+O(\alpha)$; the full expression is given by Itzykson & Zuber (1980).

The propagator (8.20) is now finite, to this order, and has a pole at $\not{p}=m$ with residue i; we write it as

$$\mathrm{i}S_{\mathrm{F}}'^{[2]}(p)=\frac{\mathrm{i}}{\not{p}-m-\Sigma_{\mathrm{R}}^{[2]}(p)} \tag{8.22}$$

(note that this is what Bjorken & Drell (1965) would call $\mathrm{i}\tilde{S}_{\mathrm{F}}'^{[2]}(p)$) where the finite ($\Lambda$-independent) quantity $\Sigma_{\mathrm{R}}^{[2]}(p)$ satisfies

$$\Sigma_{\mathrm{R}}^{[2]}(p)\big|_{\not{p}=m}=0 \tag{8.23}$$

$$\frac{\partial\Sigma_{\mathrm{R}}^{[2]}}{\partial\not{p}}(p)\bigg|_{\not{p}=m}=0. \tag{8.24}$$

We now consider the interpretation of the above procedure. We started with the QED Lagrangian (8.1), motivated from classical electromagnetic theory and the quantum-mechanical gauge principle ($p^\mu\to p^\mu-qA^\mu$). We have modified it, by the addition of (8.16) and (8.19), in order to cope with some effects due to quantum field fluctuations in the vacuum. The original Lagrangian (8.1) does not produce a finite electron propagator; the modified one

$$\begin{aligned}\mathcal{L}+\mathcal{L}_{\mathrm{ct}}&=\bar{\psi}(\mathrm{i}\not{\partial}-m)\psi+e\bar{\psi}\gamma^\mu\psi A_\mu-\tfrac{1}{4}F_{\mu\nu}F^{\mu\nu}\\&\quad-\frac{1}{2\xi}(\partial\cdot A)^2+\delta m^{[2]}(\Lambda)\bar{\psi}\psi\\&\quad+(Z_2^{[2]}(\Lambda)-1)(\bar{\psi}(\mathrm{i}\not{\partial}-m)\psi)\end{aligned} \tag{8.25}$$

does, at least to order α. Looking at (8.25), we see that the first and last terms combine to give

$$Z_2^{[2]}(\Lambda)\bar{\psi}(i\not{\partial}-m)\psi. \tag{8.26}$$

Since $\delta m^{[2]}$ is of order α, while $Z_2^{[2]}$ is $1+O(\alpha)$, we may include the $\delta m^{[2]}$ terms in (8.26) also, obtaining

$$Z_2^{[2]}(\Lambda)\bar{\psi}(i\not{\partial}-m+\delta m^{[2]}(\Lambda))\psi \tag{8.27}$$

consistently to order α. The form (8.27) has an immediate simple interpretation. We may introduce an 'unrenormalised mass' m_{un} via

$$m_{\text{un}} = m - \delta m^{[2]}(\Lambda) \tag{8.28}$$

and an 'unrenormalised field' ψ_{un} via

$$[Z_2^{[2]}(\Lambda)]^{1/2}\psi = \psi_{\text{un}}(\Lambda), \tag{8.29}$$

in terms of which (8.27) becomes

$$\bar{\psi}_{\text{un}}(\Lambda)(i\not{\partial}-m_{\text{un}}(\Lambda))\psi_{\text{un}}(\Lambda). \tag{8.30}$$

This has just the form of a free Dirac Lagrangian, for the unrenormalised field $\psi_{\text{un}}(\Lambda)$, of appropriate mass $m_{\text{un}}(\Lambda)$. The quantities are cut-off-dependent, but in just such a way that the electron propagator (at least to order α, so far) as calculated from (8.30) is finite (Λ-independent). It is, in fact, the renormalised propagator (8.22), to this order. The unrenormalised mass $m_{\text{un}}(\Lambda)$ and the unrenormalised field $\psi_{\text{un}}(\Lambda)$ are changed by the interactions - the first by the addition of $\delta m^{[2]}$, the second by a scale factor.

Returning to (8.20) and (8.22), we may rewrite the finite, renormalised propagator (retaining only terms of order α) as

$$\begin{aligned}
iS_F'(p) &= \frac{i}{\not{p}-m-\Sigma_R^{[2]}(p)} \\
&= \frac{i}{Z_2^{[2]}(\Lambda)(\not{p}-m)-\Sigma^{[2]}(p,\Lambda)+\delta m^{[2]}(\Lambda)} \\
&= \frac{i}{Z_2^{[2]}(\Lambda)(\not{p}-m_{\text{un}})+\delta m^{[2]}(\Lambda)(1-Z_2^{[2]}(\Lambda))-\Sigma^{[2]}(p,\Lambda)} \\
&\approx \frac{i}{Z_2^{[2]}(\Lambda)(\not{p}-m_{\text{un}})-\Sigma^{[2]}(p,\Lambda)} \\
&\approx \frac{i}{Z_2^{[2]}(\Lambda)(\not{p}-m_{\text{un}}-\Sigma^{[2]}(p,\Lambda))} \\
&\approx [Z_2^{[2]}(\Lambda)]^{-1}\, iS_{F\,\text{un}}'^{[2]}(p,\Lambda),
\end{aligned} \tag{8.31}$$

since $\delta m^{[2]}$, $\Sigma^{[2]}$, and $1-Z_2^{[2]}$ are each of order α. The final expression in (8.31) shows that (to this order) the propagator as calculated with the unrenormalised quantities is Z_2 times that calculated with the physical quantities and the renormalised self energy Σ_R; this is just what is expected from (8.29), since the propagator is proportional to the Fourier transform of the time-ordered product of two spinor fields.

Of course, Fig. 8.1 is by no means the only QED correction of order α. Instead of the electron propagator, we could consider the photon one. To order α, this gets modified by the addition of the process shown in Fig. 8.4. Again using Feynman gauge, the amplitude for Fig. 8.4 is (cf. (8.3))

$$\frac{-\mathrm{i}g^{\mu\rho}}{q^2}\{\mathrm{i}\Pi^{[2]}_{\rho\sigma}\}\frac{-\mathrm{i}g^{\sigma\nu}}{q^2}, \tag{8.32}$$

where

$$\mathrm{i}\Pi^{[2]}_{\rho\sigma} = -(+\mathrm{i}e)^2\int\frac{\mathrm{d}^4k}{(2\pi)^4}\,\mathrm{Tr}\left(\gamma_\rho\frac{\mathrm{i}}{\not{k}-m+\mathrm{i}\epsilon}\gamma_\sigma\frac{\mathrm{i}}{\not{k}-\not{q}-m+\mathrm{i}\epsilon}\right), \tag{8.33}$$

the extra minus sign coming from the closed fermion loop. The integral in (8.33) is certainly infinite: it appears to be quadratically divergent, from counting powers. As before, we must introduce some regularisation (cut-off). In this case, it is crucial to introduce a procedure which respects gauge-invariance, since (as we shall see shortly) we are going to use that to argue that the degree of divergence of (8.33) is actually only logarithmic. One possibility is the Pauli-Villars (1949) regularisation, in which one introduces additional spinor fields, with large mass (tending to infinity), coupled in a gauge-invariant manner to the photon. Another would be the dimensional regularisation of 't Hooft & Veltmann (1972*a*).

In Section 4.4 we saw that if, in an amplitude involving two photons, the polarisation vector of one photon was replaced by that photon's four-momentum vector, the result would be zero (see (4.71)). The process of Fig. 8.4 is such a two-photon amplitude, but of course the photons involved will, in general, not be physical (external) ones, but virtual (internal) ones. Nevertheless, it is still true that

$$q^\rho\Pi^{[2]}_{\rho\sigma} = q^\sigma\Pi^{[2]}_{\rho\sigma} = 0, \tag{8.34}$$

as is shown, for example, by Bjorken & Drell (1964), p. 154, or by Itzykson & Zuber (1980), p. 320. The relations (8.34) are further examples of Ward identities, and will be discussed as such in the following section. At this stage we anticipate that they are actually true for the full *photon self-energy (vacuum*

Fig. 8.4. Lowest-order virtual e^+e^- pair (vacuum polarisation) correction to the photon propagator.

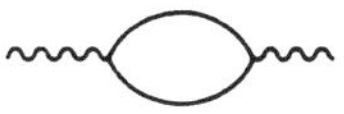

polarisation) part $\Pi_{\rho\sigma}$, to all orders, and are a consequence of gauge-invariance. (8.34) allows us to write

$$\Pi^{[2]}_{\rho\sigma}(q^2, \Lambda) = (q_\rho q_\sigma - g_{\rho\sigma} q^2)\Pi^{[2]}(q^2, \Lambda), \tag{8.35}$$

since this is the most general Lorentz covariant structure consistent with (8.34). The importance of (8.35) lies in the fact that it is the essential reason why, despite appearances, (8.33) is logarithmically - not quadratically - divergent. One can perfectly well imagine a contribution to $\Pi^{[2]}_{\rho\sigma}$ proportional to $m^2 g_{\rho\sigma}$, for example, where the m^2 is introduced on dimensional grounds, to parallel the q^2 term in (8.35). Indeed, just such a term appears in (8.33), when it is evaluated with a regularisation. The part of $\Pi^{[2]}_{\rho\sigma}$ which has the structure of (8.35) is logarithmically divergent, whereas other parts, proportional to $g_{\rho\sigma}$ alone, appear to be quadratically divergent. However, it turns out that they vanish, as the regulator masses go to infinity; for this to happen it is essential that the regularisation method be gauge-invariant. Qualitatively, one may say that the extraction of two momentum factors by (8.35) has improved the convergence of the remaining integral in $\Pi^{[2]}(q^2, \Lambda)$.

The logarithmic divergence in $\Pi^{[2]}(q^2, \Lambda)$ can be separated out, and - inserting the Λ dependence of $\Pi^{[2]}$ explicitly - one obtains (Itzykson & Zuber (1980))

$$\Pi^{[2]}(q^2, \Lambda) = \left\{\frac{\alpha}{3\pi} \ln(\Lambda^2/m^2) + \Lambda\text{-independent parts}\right\}, \tag{8.36}$$

where Λ stands for the large fermion mass(es). Just as in the case of the electron propagator, the divergence in (8.36) will be removed by the addition of a suitable counter term to the original Lagrangian (8.1). In this case, the counter term is

$$-\tfrac{1}{4}(Z_3{}^{[2]}(\Lambda) - 1)F_{\mu\nu}F^{\mu\nu}, \tag{8.37}$$

where

$$Z_3{}^{[2]}(\Lambda) = 1 - \frac{\alpha}{3\pi} \ln(\Lambda^2/m^2), \tag{8.38}$$

since (8.37) will produce an additional contribution shown in Fig. 8.5, with amplitude

$$\frac{-ig^{\mu\rho}}{q^2} \{i(Z_3{}^{[2]}(\Lambda) - 1)(q_\rho q_\sigma - q^2 g_{\rho\sigma})\} \frac{-ig^{\sigma\nu}}{q^2}. \tag{8.39}$$

Hence the contributions of Fig. 8.4 and 8.5 together produce

$$i(q_\rho q_\sigma - q^2 g_{\rho\sigma})\left\{\frac{\alpha}{3\pi} \ln(\Lambda^2/m^2) + \left(-\frac{\alpha}{3\pi}\ln(\Lambda^2/m^2)\right) + \Lambda\text{-independent parts}\right\}, \tag{8.40}$$

which is finite as $\Lambda \to \infty$.

As for the electron propagator, one can sum up the series of 'bubbles' shown in Fig. 8.6, giving

$$\mathrm{i}\tilde{D}_{\mathrm{F}}'^{[2]\mu\nu} = \mathrm{i}D_{\mathrm{F}}{}^{\mu\nu} + \mathrm{i}D_{\mathrm{F}}{}^{\mu\rho}\,\mathrm{i}\Pi_{\rho\sigma}{}^{[2]}\,\mathrm{i}D_{\mathrm{F}}{}^{\sigma\nu} + \ldots \tag{8.41}$$

or

$$\tilde{D}_{\mathrm{F}}'^{[2]\mu\nu} = D_{\mathrm{F}}{}^{\mu\nu} - D_{\mathrm{F}}{}^{\mu\rho}\,\Pi_{\rho\sigma}{}^{[2]}\tilde{D}_{\mathrm{F}}'^{[2]\sigma\nu}, \tag{8.42}$$

where $D_{\mathrm{F}}{}^{\mu\nu}(q) = -g^{\mu\nu}/q^2$, in Feynman gauge. Inserting (8.35) into (8.42) one finds

$$q^2[1 + \Pi^{[2]}(q^2, \Lambda)]\,\tilde{D}_{\mathrm{F}}'^{[2]\mu\nu} = -g^{\mu\nu} + q^{\mu}q_{\sigma}\Pi^{[2]}(q^2, \Lambda)\tilde{D}_{\mathrm{F}}'^{[2]\sigma\nu} \tag{8.43}$$

which has the solution

$$\tilde{D}_{\mathrm{F}}'^{[2]\mu\nu} = \frac{1}{q^2[1 + \Pi^{[2]}(q^2, \Lambda)]}\left[-g^{\mu\nu} - \frac{q^{\mu}q^{\nu}}{q^2}\,\Pi^{[2]}(q^2, \Lambda)\right]. \tag{8.44}$$

The counter term (8.37) is included by replacing $\Pi^{[2]}(q^2, \Lambda)$ by $\Pi^{[2]}(q^2, \Lambda) + Z_3{}^{[2]}(\Lambda) - 1$. The resulting expression is

$$D_{\mathrm{F}}'^{[2]\mu\nu} = \frac{1}{q^2[\Pi^{[2]}(q^2, \Lambda) + Z_3{}^{[2]}(\Lambda)]} \times \left(-g^{\mu\nu} - \frac{q^{\mu}q^{\nu}}{q^2}\,(\Pi^{[2]}(q^2, \Lambda) + Z_3{}^{[2]}(\Lambda) - 1)\right), \tag{8.45}$$

which we can analyse as follows. Consider the process shown in Fig. 8.7, the exchange of a photon - with all the corrections included in (8.45) - between two electrons. Since the fermion currents $\bar{u}\gamma^{\mu}u$, $\bar{u}\gamma^{\nu}u$ at either end of the photon line in Fig. 8.7 are conserved, we may disregard the $q^{\mu}q^{\nu}$ terms in (8.45). (Similar

Fig. 8.5. Contribution of the counter term (8.37) to the photon propagator.

Fig. 8.6. Iterations of Fig. 8.4.

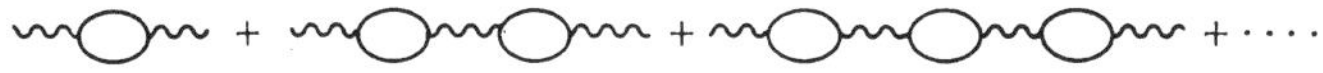

Fig. 8.7. Exchange of complete photon propagator, between two electrons.

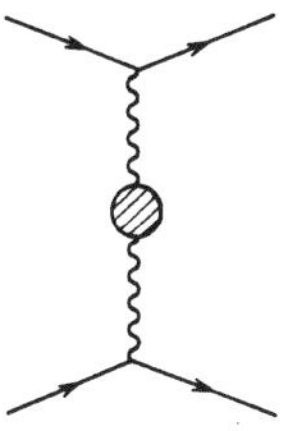

reasoning would hold whenever (8.45) was sandwiched between conserved currents.) As regards the remaining part, the position of the pole, as a function of q^2, will tell us the mass (squared) of the photon – just as in the electron case. Since $Z_3^{[2]} = 1 + O(\alpha)$, we can write this part as

$$\frac{-g^{\mu\nu}}{q^2 Z_3^{[2]}(\Lambda)(1 + \Pi^{[2]}(q^2, \Lambda))} \tag{8.46}$$

to this order, cf. (8.31). Now we have seen that if $Z_3^{[2]}$ is given by (8.38), there is no logarithmic infinity in (8.40). In fact, calculation shows also that

$$\Pi^{[2]}(q^2, \Lambda) = \frac{\alpha}{3\pi} \ln(\Lambda^2/m^2) + \text{parts tending to zero as } q^2 \to 0. \tag{8.47}$$

Thus, to this order

$$Z_3^{[2]}(\Lambda) = \{1 + \Pi^{[2]}(0, \Lambda)\}^{-1} \tag{8.48}$$

and the pole of the propagator (8.46) (renormalised through the addition of the counter term) remains at $q^2=0$ (massless photon) with unit residue. The upshot is that the renormalised photon propagator, to this order, is (up to $q^\mu q^\nu$ terms)

$$D_F'^{[2]\mu\nu}(q^2) = \frac{-ig^{\mu\nu}}{q^2[1 + \Pi_R^{[2]}(q^2)]}, \tag{8.49}$$

where $\Pi_R^{[2]}(q^2)$ is finite (Λ-independent), and tends to zero as $q^2 \to 0$. The q^2-dependent part in (8.49) implies a modification to Coulomb's law, due to vacuum polarisation effects, as first discussed by Uehling (1935); this contributes to the Lamb shift in hydrogen, as discussed in Itzykson & Zuber (1980), pp. 327-8.

We may now interpret the situation in terms of 'unrenormalised' quantities, as we did for the electron case. The propagator (8.49) will, as we said, be multiplied into current matrix elements in Fig. 8.7, each of which will involve the charge e. Thus the essential physical quantity is

$$\begin{aligned} & e^2/q^2[1 + \Pi_R^{[2]}(q^2)] \\ & \quad \approx e^2/\{Z_3^{[2]}(\Lambda)q^2[1 + \Pi^{[2]}(q^2, \Lambda)]\} \\ & \quad = e_{\text{un}}^2/q^2[1 + \Pi^{[2]}(q^2, \Lambda)] \end{aligned} \tag{8.50}$$

(cf. (8.22) and (8.31)), where

$$Z_3^{[2]}(\Lambda)e_{\text{un}}^2 = e^2. \tag{8.51}$$

On the other hand, the counter term (8.37) and the $-\frac{1}{4}F \cdot F$ term in (8.1) together yield

$$-\tfrac{1}{4}Z_3^{[2]}(\Lambda)F_{\mu\nu}F^{\mu\nu}, \tag{8.52}$$

which (cf. (8.30)) we may write as

$$-\tfrac{1}{4}F_{\text{un}\,\mu\nu}(\Lambda)F_{\text{un}}^{\mu\nu}(\Lambda) \tag{8.53}$$

if we introduce

$$[Z_3^{[2]}(\Lambda)]^{1/2}A^\mu = A_{\text{un}}^\mu(\Lambda). \tag{8.54}$$

The unrenormalised charge e_{un} is then given in terms of the renormalised (physically measured) one, e, by $e_{\text{un}} = (Z_3^{[2]}(\Lambda))^{-1/2}e$, and the corresponding fields are related by (8.54).

Actually, the question of charge renormalisation is not quite as simple as this. After all, if we look back at our initial Lagrangian, (8.1), we will realise that we have not yet discussed the renormalisation (possible addition of a counter term) for the one term in it that manifestly involves the charge! This term, $e\bar{\psi}\not{A}\psi$, gives the first-order vertex shown in Fig. 8.8, and once again we can consider the first correction to this, shown in Fig. 8.9. This time we shall simply state the answer, which is that Fig. 8.9 is logarithmically divergent, and that the divergence can be cancelled by a counter term of the form

$$e(Z_1^{[2]}(\Lambda) - 1)\bar{\psi}\not{A}\psi, \tag{8.55}$$

where $Z_1^{[2]}$ is logarithmically divergent as $\Lambda \to \infty$.

The sum of this term, and the second term in (8.1), then gives

$$eZ_1^{[2]}(\Lambda)\bar{\psi}\not{A}\psi. \tag{8.56}$$

(8.56) may be re-written in terms of ψ_{un} as

$$eZ_1^{[2]}(\Lambda)\frac{1}{Z_2^{[2]}(\Lambda)[Z_3^{[2]}(\Lambda)]^{1/2}}\bar{\psi}_{\text{un}}\not{A}_{\text{un}}\psi_{\text{un}}, \tag{8.57}$$

from which it is clear that the unrenormalised charge e_{un} is really

$$e_{\text{un}} = eZ_1^{[2]}/Z_2^{[2]}[Z_3^{[2]}]^{1/2}. \tag{8.58}$$

We shall have more to say about (8.58) in the next section.

Let us now summarise the position arrived at. We have made the *'one-loop'* diagrams of Figs 8.1, 8.4 and 8.9 finite by the addition of counter terms to the

Fig. 8.8. Lowest-order electron–photon vertex.

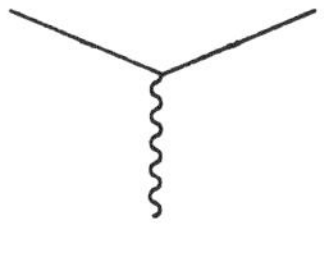

Fig. 8.9. Radiative correction to Fig. 8.8.

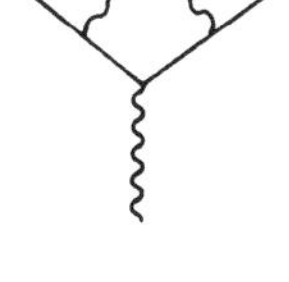

initial Lagrangian (8.1). To some extent these terms are arbitrary, and we discussed, for the electron self energy as an example, a possible set of conditions which would fix them uniquely. The counter terms could be re-interpreted as being associated with mass shifts, and scale changes on the charge and the fields. This interpretation was possible, fundamentally, because each of the counter terms that turned out to be necessary had the same *form* as one of the terms in the initial Lagrangian (8.1). No new fields or interactions had to be introduced, and consequently all that could happen (as a consequence of these second-order effects) was just these sorts of changes in the initial fields and parameters.

We may now contemplate more complicated single loop diagrams - for example, Fig. 8.10. It can be shown (Itzykson & Zuber (1980), pp. 343-4) that, as the number of external lines increases, the integral becomes more convergent, and that the only potentially dangerous diagram is actually Fig. 8.10 itself. This appears to be logarithmically divergent from power counting, and if it were really so we should have to introduce a *new* counter term of the quartic coupling ($(A \cdot A)^2$) type to cancel it. However, it turns out that - due to a Ward identity again - Fig. 8.10 is actually convergent, so no such counter term is needed. Thus the existing counter terms are sufficient to make all one-loop diagrams finite.

Finally, we may ask whether the above procedure can be extended to *all* orders in the coupling strength α. To say that a theory is renormalisable to all orders is to say precisely that the procedure can be so extended, and that all infinities are removed by the introduction of a finite number of counter terms which preserve the structure of the initial Lagrangian. This is actually true of QED, and so we may now remove all the superscript $^{[2]}$'s, and assert that constants Z_1, Z_2, Z_3 and δm can be found such that the complete Lagrangian ((8.1) plus counter terms)

$$\mathcal{L}_{\text{total}} = Z_2\bar{\psi}(\mathrm{i}\partial\!\!\!/ - m + \delta m)\psi + Z_1 e\bar{\psi}A\!\!\!/\psi - \tfrac{1}{4}Z_3F_{\mu\nu}F^{\mu\nu} - \frac{1}{2\xi}(\partial \cdot A)^2 \tag{8.59}$$

$$= \bar{\psi}_{\text{un}}(\mathrm{i}\partial\!\!\!/ - m_{\text{un}})\psi_{\text{un}} + e_{\text{un}}\bar{\psi}_{\text{un}}A\!\!\!/_{\text{un}}\psi_{\text{un}} - \tfrac{1}{4}F_{\text{un}\,\mu\nu}F_{\text{un}}{}^{\mu\nu} - \frac{1}{2\xi_{\text{un}}}(\partial \cdot A_{\text{un}})^2 \tag{8.60}$$

Fig. 8.10. One-loop graph in $\gamma\gamma \to \gamma\gamma$.

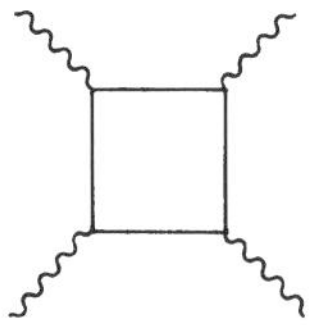

leads to finite amplitudes† for all processes. The precise form of the counter terms has to be found order-by-order in perturbation theory, by applying conditions such as (8.23) and (8.24), generalised to the removal of the $^{[2]}$'s. It is indeed quite remarkable that this assertion about (8.59) can be proved at all: a review is provided in Itzykson & Zuber, Chapter 8.

We end this section with some brief comments on technical points we have passed over. We have considered only the Feynman gauge for the uncorrected photon propagator - for example, in (8.4). We may consider instead the more general form (4.44), and $Z_2^{[2]}$ will then depend on the gauge parameter ξ. So, it turns out, will $Z_1^{[2]}$, but $Z_3^{[2]}$ and $\delta m^{[2]}$ are gauge-independent. (The latter is perhaps reasonable physically, as we should not want something with the interpretation of a *mass* to depend on the gauge.) Neither Z_1 nor Z_2 are physical quantities and - as we shall see in the following Section - they are actually equal so that their ratio disappears from (8.58), leaving only the gauge-independent quantity Z_3. The significance of this cancellation between Z_1 and Z_2 will be discussed in Section 8.2.

We may also note, in connection with questions of gauge, that it is not necessary to introduce a counter term of the gauge-fixing type. By implication, therefore, the last two terms in (8.59) and (8.60) have been equated, and we have introduced an unrenormalised gauge parameter ξ_{un} such that

$$\xi_{\text{un}} = Z_3(\Lambda)\xi. \tag{8.61}$$

We discussed the question of the photon mass in the context of the one-loop correction $\Pi^{[2]}$. Schwinger (1962) was the first to raise the possibility that the complete $\Pi(q^2)$ might - through non-perturbative effects - develop a pole at $q^2 = 0$. If it does, the photon will then become massive. We would now interpret this state of affairs in terms of a hidden gauge symmetry, the pole in $\Pi(q^2)$ at $q^2 = 0$ corresponding to the massless Goldstone quantum which couples to the photon, and shifts the mass according to the mechanism discussed in Section 6.9. Thus, if the QED gauge-invariance is not hidden, the photon mass will remain zero.

Lastly, we refer once more to the problem of infra-red divergences, and simply state that they can be handled here by the addition of a small mass to the photon. This will necessitate using a propagator of the form (6.114); the details are given in Itzykson & Zuber (1980). Z_1 and Z_2 are infra-red divergent, while δm and Z_3 are infra-red finite. These infra-red divergences disappear from physical cross-sections when the contribution of soft-photon bremsstrahlung is included (see, for example, Bjorken & Drell (1964), pp. 172-6).

† For a review of the present status of QED predictions, see Kinoshita (1979).

8.2 Ward–Takahashi identities

The generic term 'Ward–Takahashi identities' refers to identities satisfied by the amplitudes of a given theory by virtue of a *symmetry* possessed by the theory. As we have seen, a global symmetry in quantum field theory gives rise to conserved charges, and associated differentially conserved currents. Typically, Ward identities follow from the simple fact that such a symmetry current is divergenceless. The resulting identities have many uses, some of which have been alluded to in earlier chapters, for the case of QED for example. They are particularly significant for gauge theories, which are theories with symmetries *par excellence.* In this section, we shall discuss some important aspects of Ward identities.

We begin with the identity derived by Ward (1950) himself, which is used to prove the equality

$$Z_1 = Z_2 \tag{8.62}$$

between the renormalisation constants of QED. Z_1 is the vertex renormalisation constant, as given in (8.56). The identity relates the electron–photon vertex to the electron propagator, and we must first define the complete vertex. We do this by the following steps (cf. Itzykson & Zuber (1980), pp. 409-10). We consider the Lagrangian (8.1), *without* (as yet) any counter terms. We introduce the vertex V_μ by

$$\begin{aligned} &ie(2\pi)^4\delta^4(p'-p-q)\,V_\mu(p',p)\\ &\quad=\int d^4z\ d^4y\ d^4z\ \exp i(p'\cdot y-p\cdot z-q\cdot x)\\ &\qquad\times\langle 0|T(A_\mu(x)\,\psi(y)\,\bar\psi(x))|0\rangle. \end{aligned} \tag{8.63}$$

Next, observe that the vacuum expectation value of the time-ordered product will contain propagators for each of the fields, which, when Fourier transformed, will appear as factors in V_μ:

$$V_\mu(p',p) = \tilde D'_{F\mu\nu}(q)\ i\tilde S'_F(p')\Lambda^\nu(p',p)\ i\tilde S'_F(p). \tag{8.64}$$

Λ^ν is the complete electron–photon vertex; in lowest-order perturbation theory it is simply γ^μ. We recall that $\tilde S'_F$ and $\tilde D'_F$ are complete propagators (see (8.8)) without introduction of counter terms, but suitably regularised (although we shall frequently not indicate the cut-off dependence explicitly).

We now state the Ward identity:

$$\Lambda_\mu(p,p) = \frac{\partial}{\partial p^\mu}\,[\not p - m - \Sigma(p)] = \frac{\partial}{\partial p^\mu}\,[\tilde S_F'^{-1}(p)]. \tag{8.65}$$

Alternatively, setting

$$\Lambda_\mu(p',p) = \gamma_\mu + \Gamma_\mu(p',p), \tag{8.66}$$

(8.65) is

$$\Gamma_\mu(p,p) = -\frac{\partial}{\partial p^\mu}\,[\Sigma(p)]. \tag{8.67}$$

We shall discuss the proof of (8.65) or (8.67) in a moment: first we want to extract (8.62) from (8.67). When Γ_μ is calculated in perturbation theory based on (8.1), the first contribution is the diagram shown in Fig. 8.9, which has amplitude

$$\Gamma_\mu^{[2]}(p',p) = (+\mathrm{i}e)^2 \int \frac{\mathrm{d}^4 k}{(2\pi)^4}\, \frac{-\mathrm{i}}{k^2+\mathrm{i}\epsilon}\,\gamma_\nu\,\frac{\mathrm{i}}{\not p' - \not k - m + \mathrm{i}\epsilon}$$

$$\times\,\gamma_\mu\,\frac{\mathrm{i}}{\not p - \not k - m + \mathrm{i}\epsilon}\,\gamma^\nu. \tag{8.68}$$

This integral is logarithmically divergent, and the divergence is cancelled by the counter term $e(Z_1^{[2]} - 1)\bar\psi \not A \psi$ mentioned in the previous section. We can define, to all orders, the renormalised vertex $\Lambda_\mu{}^{\mathrm{R}}(p',p)$ by

$$\Lambda_\mu{}^{\mathrm{R}}(p',p) = Z_1 \Lambda_\mu(p',p) \tag{8.69}$$

or, equivalently,

$$\Gamma_\mu(p',p) = \gamma_\mu(Z_1{}^{-1} - 1) + Z_1{}^{-1}\Gamma_\mu{}^{\mathrm{R}}(p',p). \tag{8.70}$$

(8.69) is certainly plausible, from the fact that the full interaction, with the counter term, is $eZ_1\bar\psi\not A\psi$ – so that its matrix elements are Z_1-times those of $e\bar\psi\not A\psi$. The factor Z_1 can also be traced through via the relations (given in the preceding section) between renormalised and unrenormalised charge and fields. In (8.70), $\Gamma_\mu{}^{\mathrm{R}}$ is normalised such that

$$\bar u(p)\Gamma_\mu{}^{\mathrm{R}}(p,p)u(p) = 0, \tag{8.71}$$

for on-shell spinors satisfying $p^2 = m^2$. We now refer to (8.31), generalised to all orders, for the relation between the Σ appearing in (8.67) and the renormalised self energy $\Sigma_{\mathrm{R}}(p)$:

$$[\not p - m - \Sigma_{\mathrm{R}}(p)] = Z_2[\not p - m_{\mathrm{un}} - \Sigma(p,\Lambda)], \tag{8.72}$$

which can be written as

$$\Sigma(p,\Lambda) = \delta m - [Z_2{}^{-1} - 1](\not p - m) + Z_2{}^{-1}\Sigma_{\mathrm{R}}(p). \tag{8.73}$$

When we insert (8.73) and (8.70) into (8.67), and take on-shell matrix elements, we will get a relation between Z_1 and Z_2. From (8.73) we obtain

$$\frac{\partial\Sigma}{\partial p^\mu}(p,\Lambda) = -\gamma_\mu(Z_2{}^{-1} - 1) + Z_2{}^{-1}\frac{\partial\Sigma_{\mathrm{R}}}{\partial p^\mu}. \tag{8.74}$$

However, Σ^{R} satisfies the on-shell renormalisation condition (8.24) and, consequently, from (8.67), (8.70), (8.71) and (8.74), we deduce

$$Z_1 = Z_2. \tag{8.75}$$

Before proceeding to discuss the implications of (8.75) we briefly consider the proof of (8.67). In lowest order, the identity follows immediately by using the relation

$$\frac{\partial}{\partial p^\mu}(\not{p} - m)^{-1} = -\frac{1}{\not{p} - m}\gamma_\mu\frac{1}{\not{p} - m} \tag{8.76}$$

in (8.4) and comparing with (8.68). More intuitively, we might expect a simple relationship between the propagator and the vertex function, since the interaction has been introduced via the gauge-invariant substitution

$$\not{p} \to \not{p} + e\not{A}, \tag{8.77}$$

which will modify the momentum term $\not{p}$ appearing in the free particle propagator. Indeed, the observation can be extended to a graphical proof of (8.67) (Itzykson & Zuber (1980), pp. 336–7). We sketch here the main steps in a general proof based on (8.63), which exposes the role of current conservation. The vacuum value in (8.63) involves the electromagnetic field A_μ: in a graphical expansion, we will be concerned with diagrams involving one external free photon, coupling to the rest of the graph via the current j_μ. Thus

$$\begin{aligned}
&\int d^4x\, d^4y\, d^4z \exp i(\bar{p}' \cdot y - p \cdot z - q \cdot x) \\
&\quad \times \langle 0|T(A_\mu(x)\psi(y)\bar{\psi}(z))|0\rangle \\
&= -iD_{F\mu\nu}(q)\int d^4x\, d^4y\, d^4z \exp i(p' \cdot y - p \cdot z - q \cdot x) \\
&\quad \times \langle 0|T(j^\nu(x)\psi(y)\bar{\psi}(z))|0\rangle.
\end{aligned} \tag{8.78}$$

We contract (8.78) with q^μ, and use (8.63) and (8.64) on the left-hand side, giving

$$\begin{aligned}
&e(2\pi)^4\delta^4(p' - p - q)q^\mu\tilde{D}'_{F\mu\nu}(q)\tilde{S}'_F(p')\Lambda^\nu(p', p)\tilde{S}'_F(p) \\
&= q^\mu D_{F\mu\nu}(q)\int d^4x\, d^4y\, d^4z \exp i(p' \cdot y - p \cdot z - q \cdot x) \\
&\quad \times \langle 0|T(j^\nu(x)\psi(y)\bar{\psi}(z))|0\rangle.
\end{aligned} \tag{8.79}$$

However, the following relation also holds:

$$q^\mu\tilde{D}'_{F\mu\nu}(q) = q^\mu D_{F\mu\nu}(q). \tag{8.80}$$

This is equivalent to (8.34) to second order, and actually holds to all orders, as a consequence of current conservation (it is itself a 'Ward identity'! – see Itzykson & Zuber (1980), p. 408). But $q^\mu D_{F\mu\nu}$ is proportional to q_ν, so that (8.79) becomes

$$e(2\pi)^4\delta^4(p'-p-q)q_\nu \tilde{S}'_F(p')\Lambda^\nu(p',p)\tilde{S}'_F(p)$$
$$= -\mathrm{i}\int \mathrm{d}^4x\,\mathrm{d}^4y\,\mathrm{d}^4z\,\exp\mathrm{i}(p'\cdot y - p\cdot z - q\cdot x)\partial/\partial x^\nu$$
$$\times \langle 0|T(j^\nu(x)\psi(y)\bar{\psi}(z))|0\rangle, \tag{8.81}$$

where a partial integration has been used on the right-hand side.

Consider now the action of the gradient on the time-ordered product in (8.81). The point is sufficiently illustrated by the simpler example of the quantity

$$\partial_{x\nu}\langle 0|T(j^\nu(x)\phi(y))|0\rangle, \tag{8.82}$$

where ϕ is a scalar field. Since

$$T(j^\nu(x)\phi(y)) = \theta(x_0-y_0)j^\nu(x)\phi(y) + \theta(y_0-x_0)\phi(y)j^\nu(x), \tag{8.83}$$

the time derivative in the gradient will generate δ-functions when it acts on the θ-functions. Using the current conservation condition

$$\partial_{x\nu}j^\nu(x) = 0, \tag{8.84}$$

we find

$$\partial_{x\nu}\langle 0|T(j^\nu(x)\phi(y))|0\rangle = \delta(x_0-y_0)\langle 0|[j^0(x),\phi(y)]|0\rangle. \tag{8.85}$$

The generalisation to (8.81) is

$$\partial_{x\nu}\langle 0|T(j^\nu(x)\psi(y)\bar{\psi}(z))|0\rangle$$
$$= \langle 0|T\{[j^0(x),\psi(y)]\delta(x_0-y_0)\bar{\psi}(z)$$
$$+ \psi(y)[j^0(x),\bar{\psi}(z)]\delta(x_0-z_0)\}|0\rangle. \tag{8.86}$$

But the commutators in (8.86) can be evaluated from the canonical equal time commutation rules, since j^μ is simply $-e\bar{\psi}\gamma^\mu\psi$: one finds

$$[j^0(x),\psi(y)]\delta(x_0-y_0) = e\psi(x)\delta^4(x-y) \tag{8.87}$$
$$[j^0(x),\bar{\psi}(z)]\delta(x_0-z_0) = -e\bar{\psi}(x)\delta^4(x-z). \tag{8.88}$$

(8.81) therefore reduces to two terms of the form $\langle 0|T(\psi\bar{\psi})|0\rangle$, each of which is simply related to $\tilde{S}'_F$.

Specifically, inserting (8.86) with (8.87) and (8.88) into (8.81), and extracting the momentum conservation factor by translation invariance, one finds

$$\tilde{S}'_F(p',p)q_\nu\Lambda^\nu(p,p')\tilde{S}'_F(p) = \tilde{S}'_F(p) - \tilde{S}'_F(p') \tag{8.89}$$

or,

$$q_\nu\Lambda^\nu(p,p') = \tilde{S}'^{-1}_F(p') - \tilde{S}'^{-1}_F(p)$$
$$= [\not{p}' - m - \Sigma(p')] - [\not{p} - m - \Sigma(p)], \tag{8.90}$$

where $q = p' - p$. Differentiating with respect to p'^μ at $q = 0$ yields

$$\Lambda_\mu(p,p) = \frac{\partial}{\partial p^\mu}[\tilde{S}'^{-1}_F(p)]. \tag{8.91}$$

The more general relation (8.90), true for non-zero momentum transfer, was first derived by Takahashi (1957).

What is the importance of the equation $Z_1 = Z_2$? The essential point is that it means that the relationship (8.58) between the unrenormalised and renormalised charges, generalised to all orders,

$$e_{\text{un}} = eZ_1/Z_2Z_3^{1/2}, \tag{8.92}$$

in fact reads

$$e_{\text{un}} = e/Z_3^{1/2}. \tag{8.93}$$

In other words, charge renormalisation is solely due to vacuum polarisation effects, and is independent of the wavefunction (Z_2) and vertex (Z_1) renormalization factors. The latter will, in general, be different for different particles coupling to the photon ($e^-, \mu^-, p, \ldots$), but their effects *cancel out* in (8.93), leaving a *universal* renormalisation factor $Z_3^{-1/2}$ for all species: the ratio of renormalised to unrenormalised charge is independent of particle type. Hence if a set of unrenormalised charges are all equal - as is precisely the case in all gauge-type theories - the renormalised ones will be too. That is, the essential concept of 'universality of coupling', so characteristic of gauge theories, is preserved under renormalisation. (Of course, this has no bearing on the question of charge *quantisation*, already briefly mentioned in Section 7.6.) The point may also be seen very simply by noting that the replacement

$$p^\mu \to p^\mu + e_{\text{un}} A_{\text{un}}{}^\mu \tag{8.94}$$

is the *same* as

$$p^\mu \to p^\mu + eA^\mu \tag{8.95}$$

if use is made of (8.93) and of $A_{\text{un}}{}^\mu = \sqrt{Z_3} A^\mu$.

An extremely important extension of these ideas was made by Feynman & Gell-Mann (1958). The strength of the vector interaction in neutron β decay (after extraction of the Cabibbo factor $\cos\theta_C$, and radiative corrections) is found experimentally to equal the corresponding strength in the purely leptonic process of μ decay. One would naturally want to interpret this situation in terms of some symmetry between the leptonic and hadronic coupling strengths. But if this symmetry is assumed to hold for the unrenormalised strengths, strong interaction renormalisation effects, enjoyed by the hadrons but not by the leptons, would be expected to upset it (and it would be bizarre to imagine unrenormalized couplings originally unequal by just the right amount to emerge equal after renormalisation!). Such an equality between unrenormalised strengths can only be protected under renormalisation if the strengths correspond, in fact, to a conserved 'charge', and an associated divergenceless current, so that - via a Ward identity - the 'charge' renormalisation is independent of the species of particle coupled to the current. Feynman & Gell-Mann therefore postulated that the weak vector current (in strangeness-non-changing decays) was divergenceless, and that the associated weak charge (vector coupling constant) was conserved, having a universal value. These 'weak charges' act like hadronic isospin operators (for example, 'raising' a neutron into a proton in $n \to pe^-\bar{\nu}_e$). It is natural (as discussed in Section 6.8) to *identify* them with the isospin operators, when the (now universal) strength is extracted. The corresponding weak currents are then just the isospin currents of the symmetry group $SU(2)_{\text{isospin}}$. The scheme may be generalised to include strangeness-changing transitions, by regarding the relevant weak currents as symmetry currents of the flavour group $SU(3)_f$ (they will be conserved only to the extent that this is a good symmetry).

Of course, the conserved vector current (CVC) hypothesis, though it protects equality among unrenormalised couplings, does not of itself account for this equality, or universality. As we noted above, the analogous situation in QED is naturally comprehended by the gauge field idea, the universal replacement $p^\mu \to p^\mu - qA^\mu$ being made for all particles of charge q. This is the step which effectively makes the global phase invariance associated with charge conservation into a local invariance. This suggests that weak interactions (also of vector type) might be mediated by gauge fields too, associated with some *local* symmetry group or groups. The line of thought was advanced by Bludman (1958), who considered an $SU(2)_L$ gauge theory for the weak interactions only. A key contribution was made by Glashow (1961) (similar ideas were advanced by Salam & Ward (1964)), who, by enlarging the group to $SU(2)_L \times U(1)$, succeeded in including both the electromagnetic and weak interactions in a common framework. The problem of the vector meson masses was, of course, at that time unsolved. The combination of the Higgs mechanism with the Glashow gauge group resulted in the Glashow–Salam–Weinberg theory, discussed in the previous chapter.

Manipulation similar to those in (8.81)–(8.89) can be performed in many other cases. The simple result (4.71), which follows from electromagnetic current conservation and the vanishing of the commutator $[j^0(x), j^\mu(y)]$, is an example of a 'Ward-type' identity for a physical scattering amplitude. It can be used (as in Section 4.4) to eliminate unwanted photon polarisation states, and also (Itzykson & Zuber (1980), pp. 536–7) to derive a low energy theorem for Compton scattering which is independent of perturbation theory. When combined with dispersion relations, such low energy theorems lead to sum rules. Many examples are given in the book by de Alfaro et al. (1973).

It should be emphasised that, in claiming that Ward-type identities hold for physical amplitudes, a number of important questions have been begged. In particular, physical amplitudes must clearly be the *renormalised* ones. But all the manipulations done so far involve regularised amplitudes, with no inclusion of counter terms. The question of the possible impact of the regularisation procedure on the Ward identity will be taken up at the end of this section, and discussed in the following one. Assuming that the identities are unaffected by the regularisation, one must also check that the counter terms do not ruin them. This is discussed for QED by Itzykson & Zuber (1980), Section 8-4-3. The essential point is that the renormalised amplitudes are related to the unrenormalised regularised ones by multiplicative factors (the Z's), and hence the identities are preserved for the physical amplitudes.

Exactly the same procedures can also be followed for the *axial* vector currents in weak interactions. We saw in Section 6.8 that these cannot be exactly conserved unless the pseudoscalar meson masses are zero; the real world was then interpreted as an approximately chiral symmetric ($\partial \cdot j^{a\mu} = \partial \cdot j_5{}^{a\mu} = 0$) one, the axial symmetry being realised in the Goldstone mode, with the appearance of massless pseudoscalars. In the limit of exact chiral symmetry, Ward-type identities can easily be derived. As an illustration, consider the amplitude

$$T_{\mu\nu}{}^{ab} = \int d^4x \, e^{iq\cdot x} \langle p | T(j_{5\mu}{}^a(x) j_{5\nu}{}^b(0)) | p \rangle \tag{8.96}$$

involving two axial vector currents, where $|p\rangle$ is a hadron state (for example, a nucleon) of momentum p. Contracting with q^μ as in (8.81)-(8.85), we find (using $\partial^\mu j_{5\mu}{}^a(x) = 0$)

$$\begin{aligned} q^\mu T_{\mu\nu}{}^{ab} &= i \int d^4x \, e^{iq\cdot x} \partial^\mu \langle p | T(j_{5\mu}{}^a(x) j_{5\nu}{}^b(0)) | p \rangle \\ &= i \int d^4x \, e^{iq\cdot x} \delta(x_0) \langle p | [j_{50}{}^a(x), j_{5\nu}{}^b(0)] | p \rangle. \end{aligned} \tag{8.97}$$

The equal time commutator of the two currents presents some technical difficulties. It can of course be evaluated in a model, such as $j_{5\mu}{}^a \sim \bar{\psi}\gamma_\mu\gamma_5\tau^a\psi$, $j_\mu{}^a \sim \bar{\psi}\gamma_\mu\tau^a\psi$, in which case one finds

$$\delta(x_0)[j_{50}{}^a(x), j_{5\nu}{}^b(0)] = \delta^4(x) \, i\epsilon_{abc} j_\nu{}^c(0). \tag{8.98}$$

However, as already noted in Section 4.4, such current-current commutators may be model dependent: in particular, so-called Schwinger terms (which involve derivatives of δ-functions) can appear on the right-hand side of (8.98). More detailed investigation (Jackiw 1972; Itzykson & Zuber 1980, Section 5-1-7) discloses a further irregularity, having to do with the correct covariant definition of the time-ordering symbol T, when acting on a product of currents; this can lead to the appearance of additional 'seagull-terms' in $T_{\mu\nu}$. However, in cases of interest, it seems that these two subtleties cancel against each other, leaving the naive results (8.97) and (8.98) effectively valid. Alternatively, since we shall shortly proceed to consider $q \to 0$, we could take that limit at this stage in (8.97). We could then perform the spatial integral (having set $x_0 = 0$), and obtain the commutator $[Q_5{}^a(0), j_{5\nu}{}^b(0)]$, for which these difficulties do not arise; it is equal simply to $\mathrm{i}\epsilon_{abc}\, j_\nu{}^c(0)$.

Following the route of (8.97) and (8.98) we obtain

$$q^\mu T_{\mu\nu}{}^{ab} = -\epsilon_{abc} \int \mathrm{d}^4x\ \mathrm{e}^{\mathrm{i}q\cdot x} \langle p|j_\nu{}^c(0)|p\rangle \delta^4(x). \tag{8.99}$$

But the vector current is identified with the isospin current, which has matrix element

$$\langle p|j_\nu{}^c(0)|p\rangle = 2p_\nu I_\mathrm{h}{}^c, \tag{8.100}$$

where $I_\mathrm{h}{}^c$ is the hadron isospin operator (in the matrix representation appropriate to the hadron in $|p\rangle$). So finally we have the result

$$q^\mu T_{\mu\nu}{}^{ab} = -2\epsilon_{abc} p_\nu I_\mathrm{h}{}^c. \tag{8.101}$$

In this example, the commutator term has yielded a non-vanishing right-hand side in (8.101). By considering the limit $q \to 0$, and the contribution from massless pions to the left-hand side, formulae can be obtained (Tomozawa 1966; Weinberg 1966) for s-wave π-h scattering lengths. When combined with a dispersion relation, these also lead to the Adler (1965*c*)-Weissberger (1965) relation. These predictions are well obeyed (to about 10%) experimentally.

After the discussion of generalisations of the QED Ward identities associated with various global symmetries (for example, global chiral symmetry), we now turn - briefly - to their generalisation in the local non-Abelian case. The corresponding identities were first derived by Slavnov (1972) and Taylor (1971). In the case of QED, we saw how the Ward identity result $Z_1 = Z_2$ ensured that renormalisation did not spoil the universality of the gauge-invariant coupling p^μ-qA^μ. In the non-Abelian case there are complications, due to the ghost fields, which themselves become renormalised. As we saw in Section 4.6, the gauge-fixed Lagrangian with ghost fields present is invariant under the BRS symmetry (4.92)-(4.95). The Slavnov-Taylor identities can be derived by making use of *this* symmetry, as is discussed in Taylor (1978) chapter 12, or Itzykson & Zuber (1980), Section 12-4. These identities can be used to show that the universality of the gauge coupling is indeed preserved under renormalisation. Further, the identities are essential in proving both the unitarity of the theory (by eliminating unphysical longitudinal polarisation states, as discussed in Sections 4.4-4.6), and the renormalisability of the theory. There are two points to note about the latter. Firstly, renormalisability can be threatened by the high energy behaviour of the longitudinal polarisation states (as discussed for the massive vector case in Section 5.2), and thus their exorcism on grounds of unitarity is also necessary for renormalisability. Secondly, the Ward identities reduce (as in QED) the number of independent counter terms to just those that can be incorporated into the original Lagrangian by a re-definition of parameters and fields. Renormalisation is discussed in Taylor (1978), chapter 12.

The identities are also used in a formal proof of the gauge-independence of physical amplitudes. This proof is only formal, since the true nature of the severe infra-red divergences of such theories (see, for example, Doria et al. 1980; Di'Lieto et al. 1981; Andrasi et al. 1981) is still not completely understood. However, in the 'hidden' case, since the vector particles have mass, the infra-red problem disappears, and gauge-independence can be proved satisfactorily. This is, of course, vital in asserting the equivalence of the R- and U-gauges.

We now have to take up one final and important point, which will lead us on to the next section. As we mentioned earlier, we have been manipulating potentially divergent quantities, which have been regularised according to some procedure. The possibility arises that the regularisation procedure may not respect the symmetry from which (via the divergenceless condition) some Ward identity appears to be derivable. Then the 'naive' Ward identity - conjectured without regard to this spoiling of the symmetry by the regularisation - will actually be wrong. One says that an 'anomaly' is present. There are two important types of anomaly, which we shall discuss in the following two sections. The first - 'γ_5 anomalies' - occur in certain amplitudes involving axial vector currents, in which the symmetry is of the chiral type; the second type of anomaly occurs in the context of scale invariance.

8.3 γ_5 Anomalies

We do not intend to give a complete discussion of this technical (and still developing) subject. Our aim is to alert the reader to the existence of these anomalies; and to indicate how they arise, and why, in some cases, they are to be welcomed, while in others they should be eliminated.

The classic example of a γ_5 anomaly occurs in a calculation of the rate for $\pi^0 \to \gamma\gamma$, based on global chiral symmetry (realised in the hidden mode, as discussed in Section 6.8). Some successful experimental consequences of supposing that this symmetry is approximately true were mentioned in Section 6.8, and we gave an indication of the derivation of such results in Section 8.2. There are, however, three notorious *wrong* predictions of this theory, the first of which is that, in the chirally-symmetric limit, the amplitude for $\pi^0 \to \gamma\gamma$ should vanish (Veltman 1967; Sutherland 1967). In this limit, axial vector currents such as $j_{5\lambda}{}^a$ are conserved so that Ward identities of the type (4.69) can apparently be derived for amplitudes involving $j_{5\lambda}{}^a$. Unfortunately, however, as Sutherland and

Veltman pointed out, such an identity can be used to prove that $\pi^0 \to \gamma\gamma$ vanishes. It is now accepted that an anomalous term must be included, correcting this 'naive' Ward identity, with a magnitude which may be in good agreement with the observed decay rate.

Briefly, the argument for the vanishing of $\pi^0 \to \gamma\gamma$ is as follows. Consider a three-point vertex for two photons and the third (isospin) component of the axial vector current $j_{5\lambda}{}^3$, $\langle 0|j_{5\lambda}{}^3(x)|\gamma\gamma\rangle$. In momentum space this is described by an amplitude $T_{\lambda\mu\nu}\epsilon_1{}^\mu\epsilon_2{}^\nu$, where the ϵ's are the polarisation vectors of the two photons. From $\partial^\lambda j_{5\lambda}{}^3 = 0$ we expect the result

$$q^\lambda T_{\lambda\mu\nu}\epsilon_1{}^\mu\epsilon_2{}^\nu \stackrel{?}{=} 0, \tag{8.102}$$

where q is the momentum of the ingoing $j_{5\lambda}{}^3$ line. One can next enquire what intermediate states can connect $|0\rangle$ to $|\gamma\gamma\rangle$ via $j_{5\lambda}{}^3$; among these is certainly the pion, as shown in Fig. 8.11.† Clearly, this contribution involves the amplitude for $\pi^0 \to \gamma\gamma$. But it turns out that when the photons are on-shell and transverse, and as $q^2 \to 0$ (the chiral limit implies that the pion is massless, in this hidden realisation), *only* the pion pole survives on the left-hand side of (8.102). Hence $\pi^0 \to \gamma\gamma$ vanishes in this limit. Estimates (Adler 1969) of the $O(m_\pi{}^2)$ corrections to exact chiral symmetry suggest that the predicted *physical* decay rate is a good deal too small. Consequently, this result presents a problem for (approximate) chiral symmetry.

In such a situation it is often helpful to consider a detailed calculation performed within a specific model. In the present case, we want a model which exemplifies chiral symmetry in the hidden mode, so the Lagrangian of (6.87) is an obvious choice,‡ when enlarged to include electromagnetic interactions in the usual gauge-invariant way. We note that, as a consequence of the shift (6.86), the axial current (6.75) of the model acquires an extra term $v\partial^\mu\phi^a$, but it is still formally conserved despite the fact that the fermion field in (6.87) has a mass (this, of course, is because the chiral symmetry is realised in the hidden mode). By 'formally' here we mean that the result follows when the fields are treated

Fig. 8.11. Single pion pole contribution to $\langle 0|j_{5\lambda}{}^3|\gamma\gamma\rangle$.

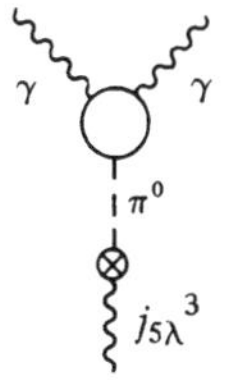

† The leptonic decay of the π involves the amplitude $\langle 0|j_{5\lambda}{}^3|\pi\rangle$.

‡ It will be sufficient to simplify (6.87) so as to include only one fermion of charge e (the proton), and two mesons, the σ and the $a = 3(\pi^0)$ component of ϕ^a.

classically. Actually, we shall see that (8.102) is *not* true in this model: an 'anomalous' term appears on the right-hand side, invalidating the classical result. The anomaly has a specifically field-theoretic origin, as is revealed by a careful study of the necessary renormalisation process.

We shall work to lowest order in the fine structure α.† We begin by checking whether the amplitude for $\pi^0 \to \gamma\gamma$ is, after all, non-vanishing in this model. To order α, there are two graphs to consider, shown in Fig. 8.12(a) and (b). The $\pi^0 \to \gamma\gamma$ amplitude has the form $\epsilon_1{}^\mu \epsilon_2{}^\nu T_{\mu\nu}$, and these two graphs give

$$T_{\mu\nu} = T_{\mu\nu}{}^{(1)}(k_1, k_2) + T_{\nu\mu}{}^{(1)}(k_2, k_1), \tag{8.103}$$

where

$$T_{\mu\nu}{}^{(1)}(k_1, k_2) = ge^2 \int \frac{\mathrm{d}^4 k}{(2\pi)^4} \times \frac{Tr[(\not{q} + \not{k} - \not{k}_1 + m)\gamma_\mu(\not{q} + \not{k} + m)\gamma_5(\not{k} + m)\gamma_\nu]}{[(q + k - k_1)^2 - m^2]\,[(q + k)^2 - m^2]\,(k^2 - m^2)} \tag{8.104}$$

and, in this model, the fermion mass is $m = -gv = gf_\pi$, as discussed in Section 6.8. The integral in (8.104) is actually convergent: the details of its evaluation are given in Itzykson & Zuber (1980), Section 11-5-2. We are interested in the limit $q^2 \to 0$, with the photons on-shell, and in this case the result is

$$\epsilon_1{}^\mu \epsilon_2{}^\nu T_{\mu\nu} \xrightarrow[q^2 \to 0]{} \frac{e^2 g}{4\pi^2 m}\, \epsilon_{\mu\nu\rho\sigma} \epsilon_1{}^\mu \epsilon_2{}^\nu k_1{}^\rho k_2{}^\sigma. \tag{8.105}$$

Using $g/m = f_\pi{}^{-1}$, the decay rate can be calculated, and good agreement with experiment is found.

We next consider the amplitude $T_{\lambda\mu\nu}$, calculated to order α in the same model, so as to shed light on the status of (8.102). The coupling of the axial

Fig. 8.12. Lowest-order (triangle) contributions to $\pi^0 \to \gamma\gamma$ in the model of Section 6.8: (a) direct; (b) crossed.

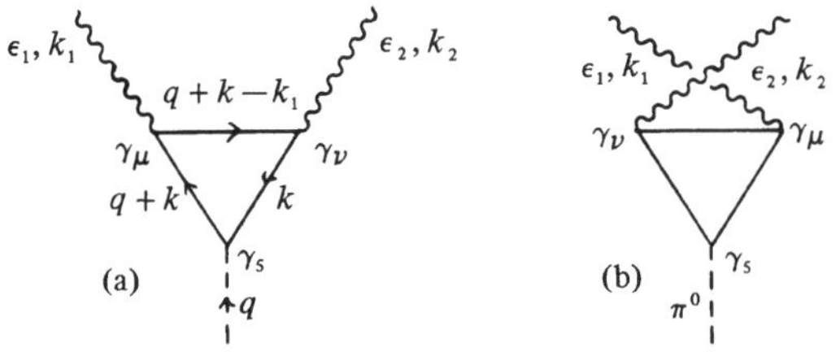

† We shall eventually see that the important results are actually true to *all* orders in α.

current can be read off by inserting (6.86) into (6.75): in addition to the ordinary axial vector coupling to the fermion, there is a direct coupling between the axial vector current and the pion, corresponding to the extra term $v\partial_\lambda\phi^a$ in $j_{5\lambda}{}^a$ which arose from the shift (6.86). The latter coupling has a factor $\mathrm{i}vq_\lambda = -\mathrm{i}(m/g)q_\lambda$. There are therefore four diagrams to consider, as shown in Fig. 8.13(a), (b), (c) and (d). We write the total contribution as

$$T_{\lambda\mu\nu} = \hat{T}_{\lambda\mu\nu} + \frac{m}{g} q_\lambda \left(\frac{-1}{q^2}\right) T_{\mu\nu}, \tag{8.106}$$

where the diagrams of Figs 8.13(a) and (b) contribute to $\hat{T}_{\lambda\mu\nu}$, and those of Figs 8.13(c) and (d) to $T_{\mu\nu}$ (the $-1/q^2$ in (8.106) comes from the massless pion pole). If (8.102) were true, therefore, we should have

$$q^\lambda \hat{T}_{\lambda\mu\nu} \stackrel{?}{=} \frac{m}{g} T_{\mu\nu}. \tag{8.107}$$

Once again, we refer to Itzykson & Zuber (1980) for the details of the calculation of the graphs in Figs 8.13(a) and (b). They show that the contribution of

Fig. 8.13. Lowest-order (triangle) contributions to the amplitude $T_{\lambda\mu\nu}$ (i.e. the axial vector$-\gamma-\gamma$ vertex): (a) and (b) axial vector coupling to fermions, direct and crossed; (c) and (d) axial vector coupling to the pion, direct and crossed.

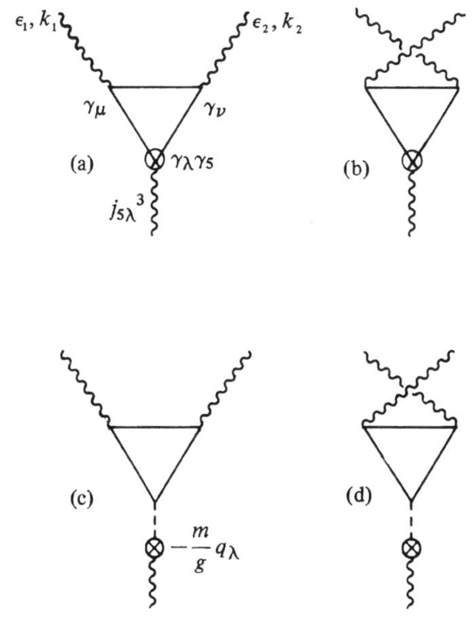

Fig. 8.13(a) to (8.107) may be written as

$$q^\lambda \hat{T}_{\lambda\mu\nu}{}^{(1)} = \frac{m}{g} T_{\mu\nu}{}^{(1)} + \frac{e^2}{2} \int \frac{d^4k}{(2\pi)^4}$$
$$\times \left\{ \frac{Tr[(\not{k} + m)\gamma_\mu\gamma_5(\not{k} - \not{k}_2 + m)\gamma_\nu]}{(k^2 - m^2)[(k - k_2)^2 - m^2]} - \frac{Tr[(\not{k} + \not{k}_1 + m)\gamma_\nu\gamma_5(\not{k} + m)\gamma_\mu]}{[(k + k_1)^2 - m^2](k^2 - m^2)} \right\}. \qquad (8.108)$$

The contribution $\hat{T}_{\lambda\nu\mu}{}^{(1)}$ of Fig. 8.13(b) is clearly the same, with $\mu \leftrightarrow \nu$, $k_1 \leftrightarrow k_2$. Thus for (8.107) to hold, the integrals in these expressions must vanish.

Here at last we arrive at the heart of the problem. We might seek to prove (8.107) by making a shift of variable $k \to k + k_2$ in (8.108). Then the first integral in (8.108) would appear to cancel against the second integral in the analogous expression for Fig. 8.13(b) (and similarly for the other two integrals). However these integrals all diverge (linearly), and the divergence invalidates such shifts of variable - which actually alter the value of the integrals by a finite amount (see Jackiw 1972). The integrals must be regularised before such shifts can consistently be performed.

At this point the question arises of what type of regularisation to adopt. We shall *assume* that electromagnetic gauge-invariance must be preserved (we comment further on this below), and use Pauli–Villars regularisation. This amounts to subtracting from every amplitude $A(m)$ depending on the fermion mass m, a corresponding amplitude $A(M)$, where the large fermion mass M will eventually tend to infinity. When these subtractions are made in (8.108), the previous argument for the vanishing of the integrals is valid, and hence we can deduce a 'regulated' form of (8.107):

$$q^\lambda [\hat{T}_{\lambda\mu\nu}(m) - \hat{T}_{\lambda\mu\nu}(M)] = \frac{m}{g} T_{\mu\nu}(m) - \frac{M}{g} T_{\mu\nu}(M). \qquad (8.109)$$

We may now let $M \to \infty$. On the left-hand side the diagrams with large fermion masses tend to zero, but on the right-hand side something remarkable happens. It is plausible (and detailed examination of (8.104) with $m \to M$ confirms this) that the $M \to \infty$ limit of $T_{\mu\nu}(M)$ is equivalent to the $q^2 \to 0$ limit. But this was already given in (8.105), and it contains a factor M^{-1}. Hence, the product $MT_{\mu\nu}(M)$ is *finite* as $M \to \infty$. Using (8.105) and (8.109), then, we obtain finally

$$q^\lambda \hat{T}_{\lambda\mu\nu} = \frac{m}{g} T_{\mu\nu} - \frac{\alpha}{\pi} \epsilon_{\mu\nu\rho\sigma} k_1{}^\rho k_2{}^\sigma. \qquad (8.110)$$

We note that, since $k_1{}^\mu \epsilon_{\mu\nu\rho\sigma} k_1{}^\rho k_2{}^\sigma = k_2{}^\nu \epsilon_{\mu\nu\rho\sigma} k_1{}^\rho k_2{}^\sigma = 0$, the electromagnetic Ward identities (following from current conservation) are preserved by (8.110).

We have discovered that, at least in this model, and with the above regularisation procedure, (8.107) is *not* correct. Neither, therefore, is (8.102) - from (8.106) and (8.110) it should read

$$q^\lambda T_{\lambda\mu\nu}\epsilon_1{}^\mu\epsilon_2{}^\mu = -\frac{\alpha}{\pi}\,\epsilon_{\mu\nu\rho\sigma}k_1{}^\rho k_2{}^\sigma\epsilon_1{}^\mu\epsilon_2{}^\nu. \tag{8.111}$$

The 'naive' Ward identity (8.102) has been corrected by the 'anomalous' contribution on the right-hand side of (8.111).

It is instructive to go back and re-examine the Sutherland-Veltman argument for the vanishing of $\pi^0 \to \gamma\gamma$, which we began by sketching. We said that, for on-shell γs and as $q^2 \to 0$, only the (massless) pion pole contributed to $q^\lambda T_{\lambda\mu\nu}\epsilon_1{}^\mu\epsilon_2{}^\nu$. In terms of the decomposition (8.106), this must mean that $q^\lambda \hat{T}_{\lambda\mu\nu}\epsilon_1{}^\mu\epsilon_2{}^\mu = 0$, as $q^2 \to 0$. Indeed this is quite true, from (8.105) and (8.110), but only by virtue of the presence of the 'anomalous' term in (8.110). For consistency, (8.102) must be corrected to (8.111) by the same anomalous piece - and this prevents the pion decay amplitude from vanishing.

Before proceeding, we comment further on the actual value predicted for the $\pi^0 \to \gamma\gamma$ decay rate. The calculation of (8.105) was of course within the model of Section 6.8 - but when the experimental parameter f_π was introduced in place of g/m, agreement with experiment was found. The fermion mass is, by itself, clearly not important. Long ago, Steinberger (1949) used a proton loop: nowadays, one would use quark loops. However, the charges and isospin-space factors, which control the strengths of the vertex couplings, are important in getting the right rate. The quark contribution to $j_{5\mu}{}^a$ is $\Sigma_i\bar{\psi}_i\gamma_\mu\gamma_5(\tau_i{}^a/2)\psi_i$, and we denote the charges by Q_i in units of e, for quarks of type i. The previous calculation is then multiplied, in the quark model, by the factor $\Sigma_i\frac{1}{2}\tau_i{}^3Q_i{}^2$. For u, d and s quarks alone, this yields a factor 1/3. Consequently agreement with experiment is lost unless there exist three (electromagnetically identical) replicas of each u, d, s quark. Colour supplies just this degeneracy, and thus the $\pi^0 \to \gamma\gamma$ rate constitutes important evidence for some such degree of freedom.

It might be thought that the results (8.110) and (8.111) are only valid to order α (although the $O(\alpha^2)$ correction would presumably be very small). In fact, however, Adler & Bardeen (1969) showed that the 'triangle' loops yield the *only* anomalous contributions for the $j_{5\lambda}{}^a$-γ-γ vertex, so that (8.110) and (8.111) are true to all orders in α.

The essential reason for the failure of the naive identity (8.102) is to be found in the fact that the regularisation adopted did not respect the (classical) symmetry on which (8.102) was based. Chiral invariance is explicitly broken by the fermion mass M in the Pauli-Villars procedure. By contrast, this regularisation, as noted above, is such that the electromagnetic Ward identities remain true, since it respects current conservation. Anomalies such as this one will occur in *all* theories with fermions coupled to both vector and axial vector currents, since it is not possible to find a regularisation scheme (for all divergent graphs) which respects both vector and axial vector symmetries (Adler 1970; Jackiw 1972). A general analysis is given by Adler & Bardeen (1969).

If, as in the $\pi^0 \to \gamma\gamma$ case, the vector currents are coupled to a gauge field, we seem to have no practical alternative to preserving the vector Ward identities, in order to ensure - as we have emphasised - the cancellation of unphysical polarisation states, and renormalisability itself. We may therefore take the view that the naive identity (8.102) is not really consistent with electromagnetic gauge-invariance (since the regularisation which respects that symmetry yields (8.111) instead of (8.102)). Alternatively, we may infer that the formal conservation law $\partial^\lambda j_{5\lambda}{}^a = 0$ is not consistent with gauge-invariance, in the presence of an electromagnetic field. This was the point of view actually taken by Schwinger (1951), who was the first to notice the existence of anomalies. In Schwinger's approach, one focusses attention on the short distance (coordinate space) singularities of operator expressions such as $\bar{\psi}(y)\gamma_\lambda\gamma_5(\tau^3/2)\psi(x)$, as $x \to y$, rather than on the high energy (momentum space) singularities of loop integrals. Such an operator is not gauge-invariant, since the local phase factors of $\bar{\psi}(y)$ and $\psi(x)$ will be different. Schwinger's analysis shows that, when it is made gauge-invariant, and the limit $x \to y$ taken, one obtains

$$\partial^\lambda j_{5\lambda}{}^3 = -\frac{\alpha}{8\pi}\epsilon_{\lambda\mu\nu\rho}F^{\lambda\mu}F^{\nu\rho}, \qquad (8.112)$$

where $F^{\lambda\mu}$ is the electromagnetic field tensor. (8.112) then modifies (8.102) to (8.111).

In the foregoing discussion the axial currents have been associated (explicitly or implicitly) with a *global* symmetry: only the vector currents have been coupled to a gauge field, with a consequent local symmetry. Thus an equation of the form (8.111) could be tolerated, since we were not coupling the axial current $j_{5\lambda}{}^3$ to a (massless) gauge field: if we had been, we should have *needed* the anomaly free result (8.102) to be true in order to ensure unitarity and renormalisability, exactly as in the vector gauge field case. In the GSW model of weak interactions, however, similar axial currents are indeed coupled to gauge fields. Here the symmetry is a non-Abelian one, and consequently the axial identities we need to preserve are the non-Abelian (Slavnov–Taylor) analogues of (8.102). Since we also need to preserve electromagnetic gauge-invariance, we must somehow ensure that no anomalies occur which involve currents coupled to gauge fields.

In the GSW model, only triangle graphs cause anomalies. As we have seen, however, the anomalous contribution is independent of the fermion mass. Thus the possibility arises of *cancelling* any anomaly present in the hadronic part of the current by the corresponding anomaly in the leptonic part of the current. Bouchiat et al. (1972) were the first to point out that *each* generation in (7.80) will be separately anomaly free if the fractionally charged quarks come in three colours. Anomaly cancellation appears to be a powerful constraint on possible theories ('t Hooft 1980).

In conclusion, we mention briefly the two other wrong predictions of chiral symmetry, whose resolution may also be related to anomalies. In the (chiral) limit, the axial *isoscalar* charge

$$Q_5{}^0 = \int \bar{\psi}\gamma_0\gamma_5\psi \, \mathrm{d}^3x \tag{8.113}$$

is conserved. In this case the Goldstone boson would presumably be the η. However, the η mass is considerably less like zero than is m_π. If m_{u} and m_{d} are allowed to be non-zero, it can be shown that $m_\eta < \sqrt{3}m_\pi$ (Weinberg 1975). This therefore poses a problem for chiral symmetry, since experimentally $m_\eta \approx 4m_\pi$. A further difficulty is that, in the chiral limit, η is forbidden to decay into 3π (Sutherland 1966). The complete resolution of these '$U(1)$ problems' is still not completely clear, although it is certainly plain that anomalies associated with the axial isoscalar current provide the starting point. Kogut & Susskind (1975) and 't Hooft (1976) have suggested possible resolutions; a recent review is provided by Peccei (1980).

8.4 Scale invariance

The study of scale invariance in quantum field theory has important applications in both particle physics and statistical mechanics. An example of the former is 'deep' inelastic lepton scattering from nucleon targets; of the latter, critical phenomena and phase transitions. In inelastic lepton scattering, a lepton of energy E in the laboratory system scatters from a nucleon, emerging with energy E', having transferred energy $E - E'$ to the nucleon. There has also been a transfer of momentum, denoted by the invariant square q^2 (see Fig. 8.14). Only the outgoing lepton is observed. Introducing $\nu = M(E - E')$, where M is the nucleon mass (ν can be written as a Lorentz invariant), the differential cross-section for this process involves (in addition to known factors) the *structure function*† $F(\nu, q^2)$, which is essentially the modulus squared of the lower vertex in Fig. 8.14, summed over the unobserved hadron final states. F is defined as a dimensionless quantity.

† Strictly speaking there are two such functions in the unpolarised case, but what we shall say applies to both, so we simplify matters by discussing a generic one.

The basic motivation for the study of scale invariance is as follows. Since F is dimensionless, it can only depend, presumably,† on dimensionless ratios ν/q^2, M_i^2/ν, M_i^2/q^2, where M_i stands for any relevant mass parameters in the process considered (for example, lepton, nucleon or quark masses). Now it is a remarkable fact that (at least to a good approximation - see below), for ν and q^2 larger than a few GeV^2, data for F at different ν and q^2 fall on a universal curve, which is a function of ν/q^2 only. This seems to be telling us that we are in a kinematic regime where the dependence on mass parameters can be ignored. Thus we are led to the idea that it might be fruitful to model the situation by studying *massless* theories - and it is for these that the (field theoretic) concept of scale invariance is relevant.

In statistical mechanics the role of the mass parameters is played by the inverse of the correlation length, ξ^{-1}. ξ is temperature-dependent, and at a critical point occurring at some temperature $T=T_C$, ξ may tend to infinity as some power: $\xi \sim (T-T_C)^{-P}$. Equivalently, the 'mass' ξ^{-1} tends to zero, so that again we are in a regime in which a massless theory may be appropriate.

For simplicity we consider the simple case of a scalar field theory. Since we shall be interested in the role of mass terms, we start with the massive theory, with Lagrangian

$$\mathcal{L}_m = \tfrac{1}{2}\partial_\mu\phi\partial^\mu\phi - \frac{g}{4!}\phi^4 - \tfrac{1}{2}m^2\phi^2 = \mathcal{L}_S + \mathcal{L}_B, \tag{8.114}$$

where $\mathcal{L}_B = -\tfrac{1}{2}m^2\phi^2$. The transformations we are concerned with are (global) scale transformations on the coordinates, which we write in the form

$$x \to x' = e^{-\alpha}x. \tag{8.115}$$

Such a transformation amounts, of course, to a re-definition of length scale (in a four-dimensional sense). However, it is important to distinguish it, at the

Fig. 8.14. One-photon exchange amplitude in inelastic e^--nucleon scattering.

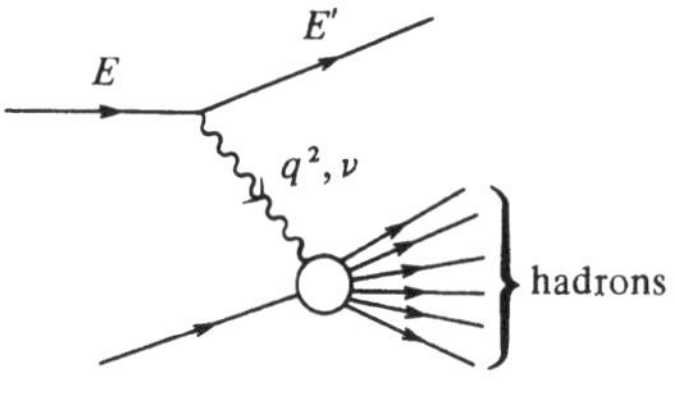

† Just this point will turn out to be *not* so obvious after all.

outset, from an ordinary dimensional rescaling of *all* quantities in (8.114) - including, in particular, the mass m. It was this kind of rescaling that we implicitly used in our dimensional analysis of F, above. Before proceeding further it is, in fact, useful to consider the behaviour of the quantities in (8.114) under such ordinary dimensional rescaling. To do this, we work in the conventional system $c = \hbar = 1$, in which only one independent unit of dimension remains - say that of mass, M. The action is then dimensionless, which means that it must be invariant under appropriate simultaneous rescaling of all dimensional quantities appearing in it. This enables us to find the dimension of the field ϕ, which will be important in what follows. Denoting dimension by square brackets, since $[I] = 0$ we have $[\mathcal{L}] = M^4$, so that from the $\partial_\mu \phi \partial^\mu \phi$ term we learn that $[\phi] = M$. We say that ϕ has dimension 1, in mass units. (In a similar way, $[g] = 0$ and, of course, $[m] = M$). Recalling that mass has the dimension of inverse length, this means that, as far as ordinary dimensional rescalings are concerned, ϕ and m are multiplied by the factor e^α when x changes by (8.115). A similar analysis shows that fermi fields have mass dimension 3/2.

However, we are actually contemplating something different from this, since the mass m in (8.114) is *not* being rescaled. We expect, then, that the term $\mathcal{L}_B$ involving m will break the invariance under the coordinate scale transformation (8.115), which is why we gave it a subscript B. To see how this occurs formally, we need to know how ϕ transforms when the coordinates change by (8.115). Since ϕ is a scalar field, it is plausible to suppose that, under (8.115), ϕ transforms into ϕ' where

$$\phi'(x') = e^{\alpha d_\phi} \phi(x); \tag{8.116}$$

that is, the new field evaluated at the new space-time point is a simple scale factor times the original field at the original space-time point. Equation (8.116) generalises the usual statement for a field which is a scalar under rotations, when the scale factor is absent. The particular form of the latter has been chosen to resemble that in (8.115); as we shall see, the parameter d_ϕ will be closely analogous to - though *not* identical with - the ordinary dimension of ϕ, in mass units.

Rewriting (8.116) as

$$\phi'(x) = e^{\alpha d_\phi} \phi(e^\alpha x) \tag{8.117}$$

it follows that the behaviour under an infinitesimal scale transformation is given by

$$\left.\begin{aligned} x &\to (1+\epsilon)x \equiv x + \delta x \\ \phi &\to \phi + \epsilon(d_\phi + x \cdot \partial)\phi \equiv \phi + \delta\phi. \end{aligned}\right\} \tag{8.118}$$

Under (8.118), we easily find

$$\left.\begin{aligned}\delta(\partial_\mu\phi\partial^\mu\phi) &= \epsilon(2 + 2d_\phi + x\cdot\partial)(\partial_\mu\phi\partial^\mu\phi)\\ \delta(\phi^4) &= \epsilon(4d_\phi + x\cdot\partial)\phi^4\\ \delta(\phi^2) &= \epsilon(2d_\phi + x\cdot\partial)\phi^2.\end{aligned}\right\} \tag{8.119}$$

In evaluating the corresponding changes in the action, an integration by parts can be performed, which converts the $x\cdot\partial$ terms into -4. We therefore obtain

$$\delta I_S \equiv \int \delta\mathcal{L}_S\, d^4x = \epsilon \int \left[\tfrac{1}{2}(2d_\phi - 2)\partial_\mu\phi\partial^\mu\phi - \frac{g}{4!}(4d_\phi - 4)\phi^4\right] d^4x \tag{8.120}$$

and

$$\delta I_B \equiv \int \delta\mathcal{L}_B\, d^4x = \epsilon \int -\tfrac{1}{2}m^2(2d_\phi - 4)\phi^2\, d^4x. \tag{8.121}$$

It follows from (8.120) and (8.121) that if $d_\phi = 1$ then the part I_S is invariant under (8.118), but I_B is not. These facts are easily interpreted. With $d_\phi = 1$, we see from (8.117) that ϕ would be rescaled by exactly the factor e^α which we obtained in our previous analysis of the ordinary dimension of ϕ; that is, the value 1 for d_ϕ is the dimension of ϕ in the sense of ordinary dimensional analysis. The discussion of the scale invariance of I goes through as before, *except* for that part of $\mathcal{L}$ which depends on the mass m, since this is not being rescaled. Thus the association of scale invariance with masslessless seems to emerge clearly.

The quantity d_ϕ, which controls the behaviour of ϕ under (8.115), is called the 'scale dimension' of the field ϕ, to distinguish it from the ordinary dimension. The reader may well wonder what the fuss is about, if indeed d_ϕ is actually equal to 1. The answer, which will emerge below, is that $d_\phi = 1$ *only in the free field case*, that is $g = 0$ in (8.114). In the interacting case, d_ϕ differs from unity by an amount which is called the 'anomalous dimension' of ϕ. Examples of anomalous dimensions in several field theories are discussed by Coleman (1973).

This fact seems to put an abrupt end to the discussion. For clearly, if $d_\phi \neq 1$, even the mass independent piece $\mathcal{L}_S$ will not lead to a scale invariant action I_S, so it would seem that, for even for massless theories, there is no such thing as scale invariance at all! Remarkably enough, however, the discussion is worth pursuing.

We shall attack this rather technical subject by easy stages. We begin with what we might again call the 'naive' approach. We certainly expect that scale invariance - in the sense of $\delta I = 0$ under (8.115) - will only stand a chance of being true if the Lagrangian contains no explicit mass terms (or if we are in a kinematic regime where masses may be neglected). Thus we consider the Lagrangian

$$\mathcal{L}_S = \tfrac{1}{2}\partial_\mu\phi\partial^\mu\phi - \frac{g}{4!}\phi^4. \tag{8.122}$$

As we have seen, $\mathcal{L}_S$ indeed leads to an invariant I_S if ϕ transforms by (8.117) with $d_\phi = 1$ (that is, in the 'naive' way). Corresponding to this symmetry, there is a Noether current, which is divergenceless (we do not give its form here - see Coleman 1973). Let us enquire what are the 'Ward identities' that follow from this condition. Consider the two-point function

$$G^{(2)}(x) = \langle 0 | T(\phi(x)\phi(0)) | 0 \rangle. \tag{8.123}$$

With $d_\phi = 1$, we would deduce that $G^{(2)}(x)$ must behave as x^{-2}; or, in momentum space, that $G^{(2)}(p) \sim p^{-2}$. Indeed, this *is* the behaviour of the *free* propagator in a massless theory (and also in a massive theory for $p^2 \gg m^2$). In general, for an n-point Green function $G^{(n)}$ with $(n-1)$ independent momenta, the scale dimension would be expected to be $n - 4(n-1)$, and thus $G^{(n)}$ would satisfy

$$\sum_{i=1}^{n-1} p_i \cdot \frac{\partial}{\partial p_i} G^{(n)} \overset{?}{=} [n - 4(n-1)]\, G^{(n)}. \tag{8.124}$$

This is typical of the 'naive' Ward identities which can be formally derived from the scale invariance of $\mathcal{L}_S$ with $d_\phi = 1$. In fact, it is false, but there is a true equation of this type (see below), which will, in appropriate circumstances, determine the x or p behaviour of the Green functions - and, indeed, of the structure functions F. The naive expectation for this behaviour, based on dimensional analysis, will turn out to be modified, but in a way which can be calculated (in perturbation theory). Herein lies the significance of the study of scale invariance, and the associated Ward identities.

The reason (8.124) is false is, once again, that the necessary regularisation conflicts with - or fails to respect - the scale invariance on which (8.124) is based. This is obviously true of a cut-off *mass*, but it is equally true of dimensional regularisation, for in the latter case it is impossible to assign (via the kinetic terms) scale dimension to the fields in such a way that an interaction term is still scale invariant, and so only free-field theories would be scale invariant - and (8.124) *is* true for the case $g = 0$, with $d_\phi = 1$. We shall discuss regularisation in terms of a cut-off Λ.

Now one might think that since Λ will in the end be taken to infinity, it will not, after all, 'set the scale', or break naive scale invariance expressed by (8.124). However, this is false. As we have seen, finite renormalised amplitudes are defined via certain *renormalisation conditions*, of the type given in (8.23) and (8.24). These are conditions on the Green functions (enough of them to determine the counter terms) at certain values of the momenta. In a massless theory we cannot choose the zero-momentum point, because of infra-red divergences. Hence, even in a massless theory, a mass scale μ will re-enter via the renormalisation conditions.

Does this mean the end of all hope of calculating the x- or p-dependence of Green functions by scale invariance arguments? It certainly cannot be done by simple scaling based on (8.124). But a more subtle and interesting situation obtains. Just as in the case of the axial anomalies considered in the previous section, careful consideration of the role of the cut-off (or of the renormalisation point μ) will produce 'anomalous' additions to the naive result (8.124). As we shall see, the origin of these terms can be understood from the seemingly very innocent fact that μ is *arbitrary*: although it has to be something, different from zero, *what* it is is irrelevant, since it is only a question of the point at which we *choose* to define our renormalised amplitudes. Our theory has to be independent of μ in the end. The next task is therefore to formulate this μ-independence of the theory.

At this point, we will be slightly more formal and state what our renormalisation conditions are for the massless theory (8.122). We introduce the proper (one-particle irreducible) Green functions $\Gamma^{(n)}$, which are just the one-particle irreducible Green function $G_{(1)}{}^{(n)}$'s with a factor $(G_{(1)}{}^{(2)})^{-1}$ for each external leg. $G_{(1)}{}^{(2)}$ is proportional to the single-particle propagator, and $\Gamma^{(2)} = (G_{(1)}{}^{(2)})^{-1}$.

A suitable set of renormalisation conditions is then

$$\Gamma^{(2)}|_{p^2=-\mu^2} = -\mu^2; \quad \left.\frac{\partial\Gamma^{(2)}}{\partial p^2}\right|_{p^2=-\mu^2} = 1; \quad \Gamma^{(4)}|_s = -g \tag{8.125}$$

where the 'symmetry' point s is $p_1{}^2 = p_2{}^2 = p_3{}^2 = p_4{}^2 = -3\mu^2/4$, $(p_1+p_2)^2 = (p_1+p_3)^2 = (p_1+p_4)^2 = -\mu^2$ (all momenta ingoing). A point in the Euclidean region is chosen, to avoid threshold singularities. The conditions (8.125) amount to a mass, wavefunction and vertex function renormalisation.

The renormalised quantities $\Gamma^{(n)}$ depend now on the momenta p_i, on g and apparently on μ. The simple scale dimension (in the sense of (8.124)) of $\Gamma^{(n)}$ is easily found to be $4-n$, so that (8.124) becomes

$$\sum_{i=1}^{n-1} p_i \cdot \frac{\partial}{\partial p_i} \Gamma^{(n)} \stackrel{?}{=} (4-n)\Gamma^{(n)}. \tag{8.126}$$

However, it is now μ that sets the scale of $\Gamma^{(n)}$, and we can write

$$\Gamma^{(n)}(p_i, g, \mu) = \mu^{4-n} f(p_i/\mu, g), \tag{8.127}$$

so that (8.126) becomes just

$$\mu \frac{\partial}{\partial \mu} \Gamma^{(n)} \stackrel{?}{=} 0. \tag{8.128}$$

We now discuss what is the true equation of the form (8.128). First, we obtain it by a formal procedure, which emphasises the process of renormalisation. $\Gamma^{(n)}$ in (8.128) is the renormalised function, and we can write it in terms of the unrenormalised one ${\Gamma_{un}}^{(n)}$ by

$$\Gamma^{(n)}(p_i, g, \mu) = Z^{n/2}{\Gamma_{un}}^{(n)}(p_i, g_{un}, \Lambda), \tag{8.129}$$

where Z is the wavefunction renormalisation constant for ϕ.

It is certainly true that

$$\frac{\partial}{\partial\mu}{\Gamma_{un}}^{(n)} = 0. \tag{8.130}$$

Hence

$$\mu\frac{\partial}{\partial\mu}(Z^{-n/2}\Gamma^{(n)}) = 0. \tag{8.131}$$

In carrying out the differentiation in (8.131), we have to pay attention to the fact that both Z and g can depend on the renormalisation point μ (g obviously does, from (8.125)). Thus (8.131) yields

$$\left(\mu\frac{\partial}{\partial\mu} + \mu\frac{dg}{d\mu}\frac{\partial}{\partial g} - \frac{n}{2}\frac{\mu}{Z}\frac{\partial Z}{\partial\mu}\right)\Gamma^{(n)} = 0 \tag{8.132}$$

or

$$\left(\mu\frac{\partial}{\partial\mu} + \beta(g)\frac{\partial}{\partial g} - \frac{n}{2}\gamma(g)\right)\Gamma^{(n)} = 0, \tag{8.133}$$

where

$$\beta(g) = \mu\frac{dg}{d\mu} \tag{8.134}$$

and

$$\gamma(g) = \mu\frac{\partial(\ln Z)}{\partial\mu}. \tag{8.135}$$

In γ, the differentiation is to be done holding g_{un} and Λ fixed. (8.133) is the equation which corrects (8.128), or (8.124). It is called a 'renormalisation group' equation.

(8.133) may be interpreted quite simply. The renormalised physical amplitude should be independent of μ. When μ changes, $\Gamma^{(n)}$ will change by (8.127), but also because g changes, and because the scale of the fields changes. For an infinitesimal change in μ, we should have

$$\Gamma\left(\mu + \delta\mu, g + \frac{dg}{d\mu}\delta\mu, \left(1 + \frac{\gamma}{2}\frac{\delta\mu}{\mu}\right)\phi\right) = \Gamma(\mu, g, \phi). \tag{8.136}$$

(8.136) is equivalent to (8.133), by a Taylor expansion.

Equation (8.133) is the condition satisfied by the $\Gamma^{(n)}$'s of a (real) scale invariant theory (such as, for example, (8.122)). β and γ can be calculated in perturbation theory. If they are known, (8.133) gives the general μ-dependence of $\Gamma^{(n)}$, and hence the p-dependence via (8.127). If both β and γ are zero, we would recover (8.128) and thence, via (8.127), the behaviour $\Gamma^{(n)} \sim (p_i^2)^{(4-n)/2}$. This is the behaviour which ordinary dimensional analysis would have given. It is called (naive) 'scaling' behaviour, because $\Gamma^{(n)}$ scales with a certain power of the momentum. Suppose, next, that only β in (8.133) is zero. Then $\Gamma^{(n)} \sim \mu^{n\gamma/2}$ and hence $\Gamma^{(n)} \sim (p_i^2)^{(4-n-n\gamma/2)/2}$. For the single-particle propagator, this situation would produce the behaviour $(p^2)^{-1+\gamma/2}$, which is scaling behaviour but with a power *different* from the simple dimensional expectation, p^{-2}. Writing this as $(p^2)^{-2+d_\phi}$, we see that the naive scale dimension $d_\phi = 1$ has been altered to $d_\phi = 1 + \gamma/2$' this is why the quantity $\gamma/2$ is called the 'anomalous dimension' of ϕ. At a zero of β, therefore, we recover scaling behaviour but with an anomalous dimension.

The solution of (8.133) is given by Coleman (1973). The important quantity is $\beta(g)$, which enters the (non-linear) differential equation

$$\frac{\mathrm{d}g}{\mathrm{d}\ln\mu} = \beta(g) \tag{8.137}$$

for the dependence of g on μ. Wilson (1971) has given a simple mechanical interpretation of such an equation which shows (as can be easily verified analytically also) that as $\mu \to \infty$, g will approach a zero of $\beta(g)$, call it g^*, which has $\mathrm{d}\beta/\mathrm{d}g|_{g^*} < 0$. From (8.133), this implies that at large (Euclidean) momenta, the $\Gamma^{(n)}$ will scale, with anomalous dimension $\gamma(g^*)$. In the infra-red limit, g approaches a zero of β which has $\beta' > 0$. Thus zeros of β control the infra-red and ultraviolet behaviour of the amplitudes.

In statistical mechanics, for the critical phenomena application, we are interested in the infra-red limit, since this is the large-distance, large correlation length, regime. Model Hamiltonians may lead to a $\beta(g^*) = 0$ with $\beta'(g^*) > 0$; the anomalous dimension $\gamma(g^*)$ is one of the 'critical indices' and has been calculated in many cases (see Amit (1978)). In particle physics, the scaling (or near scaling) behaviour observed for the structure functions implies that the naive dimensional analysis is (at least almost) true. Thus we seem to require a theory in which, in the ultraviolet regime, $g \to 0$. This in turn requires $\beta'(g=0) < 0$. It is quite remarkable that non-Abelian gauge theories have this property (Gross & Wilczek (1973*a*; Politzer 1973), which is called 'asymptotic freedom' ($g \to 0$ in the deep Euclidean region). What is more (Coleman & Gross 1973), no renormalisable theory is asymptotically free unless it contains non-Abelian gauge fields. Thus we seem forced to a non-Abelian theory (QCD) of the interquark forces.

The large momentum behaviour of an asymptotically free theory is calculable in perturbation theory, provided that there is no other zero of β with $\beta' < 0$, to which g could alternatively tend. No non-perturbative method yet exists for calculating β in general, and consequently only the vicinity of $g \approx 0$ can be analysed. Detailed calculation shows that *logarithmic* departures from exact scaling are expected in an asymptotically free theory (Gross & Wilczek 1973*b*, 1974; Georgi & Politzer 1974). A related result is that the coupling constant tends only logarithmically to zero in the large μ (or momentum) limit (the idea of a momentum dependent coupling constant was already met in the vacuum polarisation calculation in (8.49)). Exact scaling is described in terms of the parton model (pointlike non-interacting quarks - see Close 1979); the logarithmic deviations from the parton model predicted by QCD, are observed in deep inelastic lepton scattering (Perkins 1982). This constitutes strong evidence in favour of QCD.

In an asymptotically free theory, the effective coupling constant will certainly *not* end up at zero in the infra-red limit. The infra-red behaviour of QCD presents a major challenge to theory. Indeed, the large-distance aspects (confinement?) of QCD are not yet understood.

References

Abers, E. S. & Lee, B. W. (1973). *Phys. Rep.*, **9C**, no. 1.
Adler, S. L. (1965*a*). *Phys. Rev.*, **137**, B1002.
Adler, S. L. (1965*b*). *Phys. Rev.*, **139**, B1638.
Adler, S. L. (1965*c*). *Phys. Rev. Lett.*, **14**, 1051.
Adler, S. L. (1969). *Phys. Rev.*, **177**, 2426.
Adler, S. L. (1970). In *Lectures on Elementary Particles and Quantum Field Theory*, vol. 1, eds. S. Deser et al. Proceedings of the Brandeis Summer Institute (1970). Mass.: Institute of Technology.
Adler, S. L. & Dashen, R. F. (1968). *Current Algebras and Applications to Particle Physics.* New York: Benjamin.
Aitchison, I. J. R. & Hey, A. J. G. (1982). *Gauge Theories in Particle Physics.* Bristol: Adam Hilger.
Amit, D. J. (1978). *Field Theory, the Renormalisation Group and Critical Phenomena.* New York: McGraw-Hill.
Arnison, G. et al. (1983*a*). *Phys. Lett.*, **122B**, 103.
Arnison, G. et al. (1983*b*). *Phys. Lett.*, **126B**, 398.
Anderson, P. W. (1963). *Phys. Rev.*, **130**, 439.
Andraši, A., Day, M., Doria, R., Frenkel, J. & Taylor, J. C. (1981). *Nucl. Phys.*, **B182**, 104.
Aubert, J. J. et al. (1974). *Phys. Rev. Lett.*, **33**, 1404.
Augustin, J.-E. et al. (1974). *Phys. Rev. Lett.*, **33**, 1406.
Bagnaia, P. et al. (1983). *Phys. Lett.*, **129B**, 130.
Bailin, D. (1977). *Weak Interactions.* London: Chatto and Windus, for Sussex University Press.
Banner, M. et al. (1983). *Phys. Lett.*, **122B**, 476.
Barish, B. C. et al. (1975). *Phys. Rev. Lett.*, **34**, 538.
Barish, B. C. et al. (1976). *Phys. Rev. Lett.*, **36**, 939.
Barish, S.-J. et al. (1974). *Phys. Rev. Lett.*, **33**, 448.
Becchi, C., Rouet, A. & Stora, R. (1974). *Phys. Lett*, **52B**, 344.
Belavin, A. A., Polyakov, A. M., Schwartz, A. S. & Tyupkin, Y. S. (1975). *Phys. Lett.*, **59B**, 85.
Benvenuti, A. et al. (1974). *Phys. Rev. Lett.*, **32**, 800.
Benvenuti, A. et al. (1975*a*). *Phys. Rev. Lett.*, **34**, 419.
Benvenuti, A. et al. (1975*b*). *Phys. Rev. Lett.*, **35**, 1199.
Benvenuti, A. et al. (1975*c*). *Phys. Rev. Lett.*, **35**, 1203.
Bjorken, J. D. & Drell, S. D. (1964). *Relativistic Quantum Mechanics*, New York: McGraw-Hill.
Bjorken, J. D. & Drell, S. D. (1965). *Relativistic Quantum Fields*, New York: McGraw-Hill.
Bludman, S. A. (1958). *Nuovo Cimento*, **9**, 443.

Bogoliubov, N. N. (1947). *J. Phys.* (U.S.S.R.), **11**, 23.

Bogoliubov, N. N. & Shirkov, D. V. (1980). *Introduction to the Theory of Quantised Fields* (3rd edn). New York: Wiley-Interscience.

Bouchiat, C. C., Iliopoulos, J. & Meyer, Ph. (1972). *Phys. Lett.*, **38B**, 519.

Bouchiat, C. C. & Bouchiat, M. A. (1974). *Phys. Lett.*, **48B**, 111.

Boulware, D. G. & Gilbert, W. (1962). *Phys. Rev.*, **126**, 1563.

Brandt, R. (1976). *Nucl. Phys.*, **B116**, 414.

Budny, R. (1973). *Phys. Lett.*, **45B**, 340.

Buras, A. J., Ellis, J., Gaillard, M. K. & Nanopoulos, D. V. (1978). *Nucl. Phys.*, **B135**, 66.

Cahn, R. N. & Gilman, F. J. (1978). *Phys. Rev.*, **D17**, 1313.

Callan, C. G. & Treiman, S. B. (1966). *Phys. Rev. Lett.*, **16**, 153.

Chanowitz, M. S. et al. (1977). *Nucl. Phys.*, **B128**, 50.

Close, F. E. (1979). *An Introduction to Quarks and Partons.* London, New York: Academic Press.

Coleman, S. (1966). *J. Math. Phys.*, **7**, 787.

Coleman, S. (1973). In *Properties of the Fundamental Interactions*, pp. 358-99, ed. A. Zichichi. The International School of Subnuclear Physics (1971). Bologna: Editrice Compositori.

Coleman, S. & Gross, D. J. (1973). *Phys. Rev. Lett.*, **31**, 851.

Coleman, S. & Weinberg, E. (1973). *Phys. Rev.* D7, 1888.

Cummins, E. D. & Bucksbaum, P. H. (1980). *Ann. Rev. Nucl. and Part. Sci.*, **30**, 1.

Cutkosky, R. E. (1960). *J. Math. Phys.*, **1**, 429.

de Alfaro, V., Fubini, S., Furlan, G. & Rosetti, C. (1973). *Currents in Hadron Physics.* Amsterdam: North-Holland.

Di'Lieto, C., Gendron, S., Halliday, I. G. & Sachrajda, C. T. (1981). *Nucl. Phys.*, **B183**, 223.

Dokshitzer, Yu. L., Dyakonov, D. I. & Troyan, S. I. (1980). *Phys. Rep.*, **58**, 269.

Doria, R., Frenkel, J. & Taylor, J. C. (1980). *Nucl. Phys.*, **B168**, 93.

Ellis, J. (1977). In *Weak and Electromagnetic Interactions at High Energy*, eds. R. Balian & C. H. Llewellyn Smith. North-Holland.

Ellis, J. & Sachrajda, C. (1980). CERN TH 2782.

Englert, F. & Brout, R. (1964). *Phys. Rev. Lett.*, **13**, 321.

Fabri, E. & Picasso, L. E. (1966). *Phys. Rev. Lett.*, **16**, 408.

Faddeev, L. D. & Popov, V. N. (1967). *Phys. Lett.*, **25B**, 29.

Feldman, G. J. et al. (1977). *Phys. Rev. Lett.*, **38**, 1313.

Fermi, E. (1934*a*). *Nuovo Cimento*, **11**, 1.

Fermi, E. (1934*b*). *Z. Phys.*, **88**, 161.

Fetter, A. L. & Walecka, J. D. (1971). *Quantum Theory of Many-Particle Systems.* New York: McGraw-Hill.

Feynman, R. P. (1963). *Acta Physica Polonica*, **26**, 697.

Feynman, R. P. & Gell-Mann, M. (1958). *Phys. Rev.*, **109**, 193.

Fock, V. (1927). *Z. Phys.*, **39**, 226.

Fortson, E. N. & Wilets, L. (1980). *Adv. in At. and Mol. Phys.*, **16**, 319.

Gaillard, M. K. & Lee, B. W. (1974). *Phys. Rev.*, D **10**, 897.

Gastmans, R. (1975). In *Weak and Electromagnetic Interaction at High Energies*, eds. M. Levy, J.-L. Basdevant, D. Speiser & R. Gastmans. The Cargèse School (1975). New York, London: Plenum.

Gell-Mann, M. (1961). *Phys. Rev.*, **125**, 1067.

Gell-Mann, M. & Levy, M. (1960). *Nuovo Cimento*, **16**, 705.

Georgi, H. & Glashow, S. L. (1972). *Phys. Rev. Lett.*, **28**, 1494.

Georgi, H. & Glashow, S. L. (1974). *Phys. Rev. Lett.*, **32**, 438.

Georgi, H. & Politzer, H. D. (1974). *Phys. Rev.*, D**9**, 416.
Georgi, H., Quinn, H. R. & Weinberg, S. (1974). *Phys. Rev. Lett.*, **33**, 451.
Glashow, S. L. (1961). *Nucl. Phys.*, **22**, 579.
Glashow, S. L. & Gell-Mann, M. (1961). *Ann. Phys.*, **15**, 437.
Glashow, S. L., Iliopoulos, J. & Maiani, L. (1970). *Phys. Rev.*, D**2**, 1285.
Goldberger, M. & Treiman, S. B. (1958). *Phys. Rev.*, **110**, 1178.
Goldhaber, G. et al. (1976). *Phys. Rev. Lett.*, **37**, 255.
Goldman, T. & Ross, D. A. (1979). *Phys. Lett.*, **84**B, 208.
Goldman, T. & Ross, D. A. (1980). *Nucl. Phys.*, B**171**, 273.
Goldstone, J. (1961). *Nuovo Cimento*, **19**, 154.
Goldstone, J. Salam, A. & Weinberg, S. (1962). *Phys. Rev.*, **127**, 965.
Gross, D. J. (1974). *Phys. Rev. Lett.*, **32**, 1071.
Gross, D. J. & Wilczek, F. (1973*a*). *Phys. Rev. Lett.*, **30**, 1343.
Gross, D. J. & Wilczek, F. (1973*b*). *Phys. Rev.*, D**8**, 3633.
Gross, D. J. & Wilczek, F. (1974). *Phys. Rev.*, D**9**, 980.
Guralnik, G. S., Hagen, C. R. & Kibble, T. W. B. (1964). *Phys. Rev. Lett.*, **13**, 585.
Guralnik, G. S., Hagen, C. R. & Kibble, T. W. B. (1968). In *Advances in Particle Physics*, vol. 2, pp. 567ff, eds. R. Cool & R. E. Marshak. New York, London: Interscience.
Hasert, F. J. et al. (1973). *Phys. Lett.*, B**46**, 138.
Higgs, P. W. (1964*a*). *Phys. Lett.*, **12**, 132.
Higgs, P. W. (1964*b*). *Phys. Rev. Lett.*, **13**, 508.
Isgur, N. & Karl, G. (1979). *Phys. Rev.*, D**19**, 2653.
Itzykson, C. & Zuber, J.-B. (1980). *Quantum Field Theory*. New York: McGraw-Hill.
Jackiw, R. (1972). In *Lectures in Current Algebra and its Applications*, pp. 97–254, eds. S. B. Treiman, R. Jackiw & D. J. Gross. Princeton University Press.
Kamefuchi, S. (1960). *Nucl. Phys.*, **18**, 691.
Kibble, T. W. B. (1966). *Proc. Oxford Conf. on Elementary Particles 1965*, pp. 19ff. RHEL.
Kibble, T. W. B. (1967). *Phys. Rev.*, **155**, 1554.
Kim, J. E. et al. (1981). *Rev. Mod. Phys.*, **53**, 211.
Kinoshita, T. (1979). In *Proc. 19th Int. Conf. on High Energy Physics, Tokyo, 1978*, eds. S. Homma, M. Kawaguchi & H. Miyagawa. Tokyo: The Physical Society of Japan.
Kogut, J. & Susskind, L. (1975). *Phys. Rev.* D**11**, 3594.
Kugo, T. & Ojima, I. (1979). *Prog. Theor. Phys. Suppl.*, **66**, 1.
Landau, L. D. (1937*a*). *J.E.T.P.*, **7**, 19.
Landau, L. D. (1937*b*). *Phys. Z. Sowjet.*, **11**, 26, 545.
Lautrup, B. (1967). *Kon. Dan. Vid. Selsk. Mat.-Fys. Med.*, **35** (11), 1.
Lee, W. et al. (1977). *Phys. Rev. Lett.*, **38**, 202.
Levy, M. (1979). In *Hadron Structure and Lepton-Hadron Interactions*, pp. 465-502, eds. M. Levy, J.-L. Basdevant, D. Speiser, J. Weyers, R. Gastmans & J. Zinn-Justin. The Cargèse School (1977). New York, London: Plenum.
Llewellyn Smith, C. H. (1974). In *Phenomenology of Particles at High Energies*, eds. R. L. Crawford & R. Jennings. Scottish Universities Summer School (1973). London, New York: Academic Press.
Llewellyn Smith, C. H. (1980). In *Quantum Flavordynamics, Quantum Chromodynamics and Unified Theories*, eds. K. T. Mahanthappa & J. Panda. The Boulder Summer Institute (1979). New York: Plenum.
Llewellyn Smith, C. H. (1982*a*). *Phil. Trans. R. Soc. Lond.*, A**304**, 5.

Llewellyn Smith, C. H. (1982*b*). In *Rencontre de Moriond, 1981*.
Llewellyn Smith, C. H., Ross, G. G. & Wheater, J. F. (1981). *Nucl. Phys.*, **B177**, 263.
Llewellyn Smith, C. H. & Wheater, J. F. (1981). *Phys. Lett.*, **105B**, 486.
London, F. (1927). *Z. Phys.*, **42**, 375.
Maiani, L. (1976). *Proc. 1976 CERN School of Physics*, pp. 23–56. Geneva: CERN 76-20.
Mandl, F. (1959). *Introduction to Quantum Field Theory*. New York: Wiley-Interscience.
Marciano, W. J. & Sirlin, A. (1981). *Nucl. Phys.*, **B159**, 442.
Marshak, R. E., Riazuddin, & Ryan, C. R. (1969). *Theory of Weak Interactions*. New York: Wiley-Interscience.
Maskawa, K. & Kobayashi, M. (1973). *Prog. Theor. Phys.*, **49**, 652.
Matthews, P. T. (1949). *Phys. Rev.*, **76**, 1254.
Mohapatra, R. N. (1978). *Proc. 19th Int. Conf. on High Energy Physics, Tokyo, 1978*, p. 604, eds. S. Homma, M. Kawaguchi & H. Miyazawa. Tokyo: The Physical Society of Japan.
Nakanashi, N. (1966). *Prog. Theor. Phys.*, **35**, 111.
Nakanashi, N. (1973). *Prog. Theor. Phys.*, **49**, 640.
Nakanashi, N. (1974). *Prog. Theor. Phys.*, **52**, 1929.
Nambu, Y. (1960). *Phys. Rev. Lett.*, **4**, 380.
Nambu, Y. & Lurie, D. (1962). *Phys. Rev.*, **125**, 1429.
Nanopoulos, D. V. & Ross, D. A. (1979). *Nucl. Phys.*, **B157**, 273.
Oppenheimer, J. R. (1930). *Phys. Rev.*, **35**, 461.
Pauli, W. & Villars, F. (1949). *Rev. Mod. Phys.*, **21**, 434.
Peccei, R. D. (1980). In *Particle Physics 1980*, eds. L. Andrie et al. Amsterdam: North Holland.
Perkins, D. H. (1982). *Phil. Trans. Roy. Soc. Lond.*, **A304**, 23.
Perl, M. et al. (1976). *Phys. Lett.*, **63B**, 466.
Peruzzi, I. et al. (1976). *Phys. Rev. Lett.*, **37**, 569.
Politzer, H. D. (1973). *Phys. Rev. Lett.*, **30**, 1346.
Prescott, C. Y. et al. (1978). *Phys. Lett.*, **77B**, 347.
Ramond, P. (1981). *Field Theory*. Reading, Mass.: Benjamin.
Ross, D. A. (1978). *Nucl. Phys.*, **B140**, 1.
Sakurai, J. J. (1960). *Ann. Phys.*, **11**, 1.
Salam, A. (1960). *Nucl. Phys.*, **18**, 681.
Salam, A. (1968). In *Elementary Particle Theory*, p. 367, ed. N. Svartholm. Stockholm: Almquist Forlag AB.
Salam, A. & Ward, J. C. (1964). *Phys. Lett.*, **13**, 168.
Schwinger, J. (1951). *Phys. Rev.*, **82**, 664.
Schwinger, J. (1957). *Ann. Phys.* (N.Y.), **2**, 407.
Schwinger, J. (ed.) (1958). *Quantum Electrodynamics (Selected Papers on)*. New York: Dover.
Schwinger, J. (1962). *Phys. Rev.*, **125**, 397.
Sirlin, A. (1980). *Phys. Rev.*. **D22**, 971.
Slavnov, A. (1972). *Theor. and Math. Phys.*, **10**, 99.
Steinberger, J. (1949). *Phys. Rev.*, **76**, 1180.
Stueckelberg, E. C. G. (1938). *Helv. Phys. Acta*, **11**, 225.
Sutherland, D. G. (1966). *Phys. Lett.*, **23**, 384.
Sutherland, D. G. (1967). *Nucl. Phys.*, **B2**, 433.
Takahashi, Y. (1957). *Nuovo Cimento*, **6**, 370.
Taylor, J. C. (1958). *Phys. Rev.*, **110**, 1216.

Taylor, J. C. (1971). *Nucl. Phys.*, B**33**, 436.
Taylor, J. C. (1978). *Gauge Theories of Weak Interactions.* Cambridge University Press.
Tilley, D. R. & Tilley, J. (1974). *Superfluidity and Superconductivity.* New York: Van Nostrand Reinhold.
Tomozawa, Y. (1966). *Nuovo Cimento,* **46**A, 707.
't Hooft, G. (1971*a*). *Nucl. Phys.*, B**33**, 173.
't Hooft, G. (1971*b*). *Nucl. Phys.*, B**35**, 167.
't Hooft, G. (1971*c*). *Nucl. Phys.*, **37**B, 195.
't Hooft, G. (1976). *Phys. Rev.*, D**14**, 3432.
't Hooft, G. (1978). *Phys. Rev.*, D**18**, 2199.
't Hooft, G. (1980). In *Recent Developments in Gauge Theories,* eds. G. 't Hooft et al. The Cargèse School (1979). New York, London: Plenum.
't Hooft, G. & Veltman, M. (1972*a*). *Nucl. Phys.*, B**44**, 189.
't Hooft, G. & Veltman, M. (1972*b*). *Nucl. Phys.*, B**50**, 318.
Utiyama, R. (1956). *Phys. Rev.*, **101**, 1597.
Veltman, M. (1967). *Proc. Roy. Soc.*, **A301**, 107.
Waller, I. (1930*a*). *Z. Phys.*, **59**, 168.
Waller, I. (1930*b*). *Z. Phys.*, **62**, 673.
Ward, J. C. (1950). *Phys. Rev.*, **78**, 1824.
Weinberg, S. (1966). *Phys. Rev. Lett.*, **17**, 616.
Weinberg, S. (1967). *Phys. Rev. Lett.*, **19**, 1264.
Weinberg, S. (1975). *Phys. Rev.*, **D11**, 3583.
Weinberg, S. (1979). *Phys. Rev. Lett.*, **43**, 1566.
Weinberg, S. (1980). *Rev. Mod. Phys.*, **52**, 515.
Weinberg, S. (1981). *Sci. American,* **244** (6), 52.
Weisberger, W. (1965). *Phys. Rev. Lett.*, **14**, 1047.
Weyl, H. (1929). *Z. Phys.*, **56**, 330.
Wilczek, F. & Zee, A. (1979). *Phys. Rev. Lett.*, **43**, 1571.
Wilson, K. G. (1971). *Phys. Rev.*, **B4**, 3174.
Yang, C. N. (1977). *Ann. N.Y. Acad. Sci.*, **294**, 86.
Yang, C. N. & Mills, R. L. (1954). *Phys. Rev.*, **96**, 191.

Index

Printed in Great Britain
by Amazon

87231668R00109